World's First **Antibiotic**

THE ULTIMATE GUIDE TO PENICILLIN

A Remarkable Cure for Infections, Sepsis, Pneumonia, Rheumatic Fever, Bacterial Meningitis, and More, Revolutionizing Modern Medicine

Jean Lizah

Copyright © [2024] by [Jean Lizah]

All rights reserved. No part of this book may be reproduced, distributed, or transmitted in any form or by any means, including photocopying, recording, or other electronic or mechanical methods, without the prior written permission of the publisher, except in the case of brief quotations embodied in critical reviews and certain other noncommercial uses permitted by copyright law.

Disclaimer

This book is intended for educational and informational purposes only. The content within is based on the author's research, personal knowledge, and experience. It should not be considered a substitute for professional medical advice, diagnosis, or treatment. Always seek the advice of your physician or other qualified healthcare providers with any questions you may have regarding a medical condition or treatment.

The author and publisher disclaim any liability arising directly or indirectly from the use or misuse of the information provided in this book. While every effort has been made to ensure the accuracy of the information, the author and publisher do not guarantee the completeness or relevance of the material and shall not be liable for any damages or negative consequences from any action taken by a reader.

Table of Contents

Chapter 1 .... **28**
How Penicillin Works **28**
    1.1 The Science Behind Penicillin 32
    1.2 Mechanism of Action: How It Kills Bacteria 36
    1.3 Types of Bacteria Penicillin Targets 40
    1.4 Penicillin vs. Other Antibiotics 44

Chapter 2 .... **50**
Medical Uses of Penicillin **50**
    2.1 Common Infections Treated by Penicillin 55
    2.2 Different Types of Penicillin 59
    2.3 Dosage Guidelines and Administration 63
    2.4 Special Considerations: Children, Elderly, and Pregnant Women 67

Chapter 3 .... **72**
The Impact of Penicillin on Global Health **72**
    3.1 Penicillin in WWII and Public Health Campaigns 76
    3.2 Reduction in Mortality Rates for Infectious Diseases 81
    3.3 Penicillin's Role in Modern Healthcare Systems 85

Chapter 4 .... **90**
Penicillin Resistance **90**
    4.1 Understanding Antibiotic Resistance 94
    4.2 How Bacteria Develop Resistance to Penicillin .. 99
    4.3 The Rise of Resistant Strains: MRSA and More..

103

4.4 Strategies to Combat Antibiotic Resistance..107

Chapter 5..112

Side Effects and Risks..112

5.1 Common Side Effects of Penicillin....................116

5.2 Allergic Reactions: Symptoms and Management..120

5.3 Who Should Avoid Penicillin?........................... 124

5.4 Interactions with Other Medications.............. 128

Chapter 6..133

Penicillin in Modern Medicine............................133

6.1 New Applications and Combinations...............137

6.2 Penicillin in Veterinary Medicine.....................142

6.3 Alternatives When Penicillin Is Not Effective.....147

Chapter 7..152

Production and Availability..................................152

7.1 How Penicillin Is Produced Today.....................157

7.2 Challenges in Supply and Distribution............162

7.3 Access to Penicillin in Developing Countries166

Chapter 8.. 171

Controversies and Ethical Considerations............... 171

8.1 Overuse of Antibiotics and the Consequences...176

8.2 Ethical Issues in Antibiotic Distribution........180

8.3 Balancing Treatment with Prevention of Resistance..184

Chapter 9..190

The Future of Penicillin..190

9.1 Innovations in Antibiotic Research.................. 195

9.2 Penicillin Derivatives and Their Uses............ 200
9.3 Global Initiatives to Preserve Antibiotic Effectiveness..204
Conclusion... 209
Summary of Penicillin's Legacy.............................212
Moving Forward: Lessons Learned from Penicillin. 214
Encouraging Responsible Use of Antibiotics.......218
Case Studies.. 225
Glossary of Terms..229
Frequently Asked Questions (FAQ)......................233

INTRODUCTION

Penicillin is often called the "miracle drug" that changed the course of history. Before its discovery, even a small cut could become life-threatening if it got infected. The world was a different place back then; common infections had the power to turn into deadly ailments, and medical practitioners had little more than hope and luck in their arsenal. But in 1928, a chain of events set off by a curious scientist, an unexpected mold, and perhaps a bit of serendipity would transform the landscape of modern medicine forever.

Penicillin, quite literally, began as a happy accident. Alexander Fleming, a Scottish bacteriologist, had returned to his cluttered lab after a vacation, only to find that one of his petri dishes, which had been left exposed, was now growing mold. What would have been thrown away by most scientists caught Fleming's attention instead. He noticed something remarkable: the bacteria growing on the dish had been stopped in their tracks near the mold. The mold, later identified as *Penicillium notatum*, seemed to possess some kind of mysterious bacterial repelling power. It was in this moment of observation—of connecting an ordinary mishap with a bigger idea—that the world's first antibiotic was born.

It wasn't easy to bring penicillin from Fleming's petri dish to the public. For a decade, it lingered as an interesting lab experiment, but it wasn't until the early 1940s that researchers like Howard Florey, Ernst Boris Chain, and their team picked up Fleming's findings and turned penicillin into a usable medicine. Imagine the urgency of the times: World War II was raging, and soldiers were dying not just from battle wounds but from infections that followed. The race to mass-produce penicillin was as much about winning the war against infections as it was about saving lives on the battlefield. By the end of the war, penicillin had earned its nickname, the "wonder drug," and had already saved countless soldiers' lives.

Penicillin marked a revolution in medicine. Suddenly, doctors had a tool that worked—reliably and consistently—against infections that had once been considered a death sentence. Strep throat, pneumonia, syphilis, and even infected wounds that once seemed hopeless could be treated effectively. It wasn't magic; it was science in action, a testament to human ingenuity and the power of curiosity.

But penicillin wasn't just a story of laboratories and scientific discovery; it was a story of humanity. When penicillin production was scaled up, the impact rippled across the world. For families who had lost loved ones to simple infections, penicillin was a blessing. For communities that had lived under the shadow of untreatable diseases, it represented hope. The availability of penicillin turned hospitals from places of convalescence and prayer into genuine centers of healing.

And like all good stories, penicillin's story also comes with its complexities. With widespread use came challenges, including allergic reactions and the eventual rise of antibiotic resistance. But these challenges do not diminish penicillin's initial triumph; rather, they remind us of how interconnected medicine, humanity, and responsibility are.

Penicillin's journey, from a moldy petri dish in a forgotten corner of a lab to a life-saving medicine that shaped the world, is a story of more than just a drug. It's about the power of observation, the importance of perseverance, and the unexpected ways that science can change the world. Today, while there are many antibiotics available, penicillin stands as the cornerstone of modern medicine—a humble beginning that sparked a revolution.

What is Penicillin?

What exactly is penicillin? It's a question that seems simple, but the answer has shaped the way we live today. Imagine a substance that can take down dangerous bacteria, save lives, and turn potentially fatal infections into something manageable with just a few doses. That's penicillin—a groundbreaking antibiotic that changed medical history forever.

Penicillin is derived from a type of mold. That's right, the very stuff we might find growing on a forgotten slice of bread had, within it, the potential to become one of the most powerful medicines in the world. To get technical for a moment, penicillin is part of a group of antibiotics known as beta-lactams. These antibiotics work by interfering with the bacteria's ability to build their cell walls. Without a strong cell wall, the bacteria essentially fall apart—unable to defend themselves, they die off. This mechanism made penicillin the go-to weapon against many bacterial infections that once had no effective treatment.

When we think of penicillin, we can't forget Alexander Fleming's pivotal role. His discovery of penicillin in 1928 wasn't exactly planned. The story goes that he left his petri dishes unattended while he went on vacation, and when he returned, he noticed something strange. A particular mold seemed to repel the bacteria he had been studying. It was a simple observation, but one that carried immense weight. Fleming had stumbled upon something revolutionary: a substance that could kill harmful bacteria without harming human cells.

The name "penicillin" comes from the mold itself—*Penicillium notatum*. It's kind of poetic, in a way, that something so seemingly insignificant and ordinary could contain such life-saving potential. Picture Fleming, standing in his messy lab, staring at that mold and realizing the potential impact it could have. He could have easily ignored it or thrown away the contaminated dish. But instead, he pursued it. It's a reminder of how often the small things we overlook can hold the answers to big problems.

Penicillin itself isn't a single substance but rather a family of antibiotics. There are different types of penicillins, each tailored for specific kinds of bacterial infections. Penicillin G, for instance, was the original form used for injection, while Penicillin V was developed for easier oral administration. Over time, chemists have also modified penicillin to create derivatives that can tackle a broader range of bacteria or resist certain types of bacterial defenses. This versatility made penicillin a cornerstone of antibiotic therapy.

But what does all this mean on a human level? Before penicillin, even something as mundane as a scratch could lead to a life-threatening infection. People often feared minor injuries, knowing that a seemingly small cut could spiral into something deadly. Childbirth, surgeries, even dental work—all carried high risks of infection. The introduction of penicillin was like turning on a light in a dark room; suddenly, there was a clear path to treatment.

Penicillin's effectiveness wasn't just a scientific victory; it was a deeply personal one for the millions of people whose lives were saved. During World War II, for example, penicillin was given to wounded soldiers, and the results were extraordinary. Soldiers who would have died from infected wounds were now surviving and returning home. Stories of penicillin's impact spread like wildfire, and it didn't take long for the drug to become known as a miracle of modern medicine.

However, like all good stories, there's more to penicillin than just triumph. It worked wonders, but it wasn't perfect for everyone. Some people found out the hard way that they were allergic to penicillin. A sudden rash, difficulty breathing—these reactions could be frightening, and for some, even dangerous. This discovery led doctors to become more careful and attentive when prescribing it, ensuring that the wonder drug didn't turn into a hazard for those who were sensitive to it.

Today, penicillin still plays a vital role in healthcare, though it has been joined by many other antibiotics. It's a reminder of how far we've come, from the days when a simple infection could mean a death sentence, to an era where we have a powerful arsenal of tools to fight disease. Penicillin opened the door, and the medical world stepped through, forever changed. It's not just a drug—it's a symbol of hope, a testament to the importance of curiosity, and proof that even the most unexpected discoveries can lead to profound change.

The Discovery: Alexander Fleming's Breakthrough

The story of penicillin begins with a moment of pure, serendipitous genius—one of those rare instances where luck, curiosity, and science all collide to create something extraordinary. It was 1928, and a Scottish bacteriologist named Alexander Fleming was working in his cluttered, modest laboratory at St. Mary's Hospital in London. Fleming had a habit of being a bit untidy, which, as fate would have it, would lead to one of the most important medical discoveries in human history.

Fleming had been studying *Staphylococcus* bacteria—tiny, stubborn microbes that can cause everything from mild skin infections to life-threatening diseases. His lab was filled with petri dishes, some neatly stacked, others left in various corners of the room. One day, as he returned to his lab after a vacation, he noticed something curious. A petri dish he had accidentally left out had become contaminated by a mold. But there was something more to it: around the mold, there was a clear zone where the bacteria had vanished. The mold was not just growing; it seemed to be killing off the bacteria.

Most people would have thrown away the contaminated dish without a second thought. But Fleming was different. He didn't see an inconvenience—he saw a mystery worth solving. What was this mold, and why were the bacteria disappearing around it? Fleming had a hunch that this could be something important, and he began investigating further. He identified the mold as belonging to the *Penicillium* genus and realized that it was producing a substance capable of killing certain types of harmful bacteria.

Fleming named the substance "penicillin," and he quickly understood its potential to change medicine. But Fleming was a scientist, not a pharmaceutical developer. He struggled to extract penicillin in quantities large enough to experiment with, and the substance was unstable, breaking down before it could be used effectively. Though Fleming wrote about his findings and shared them with the scientific community, the world didn't immediately recognize the full implications of what he had discovered. In fact, his work sat on the back burner for nearly a decade.

It wasn't until a team of scientists led by Howard Florey, Ernst Boris Chain, and Norman Heatley picked up where Fleming left off that penicillin truly began its journey to becoming the first widely used antibiotic. Imagine the determination of these researchers as they worked tirelessly, even during World War II, to figure out how to mass-produce this life-saving substance. They experimented with every possible method—from using bedpans to grow the mold to borrowing milk bottles for production. Eventually, they succeeded, and the results were nothing short of miraculous.

For Alexander Fleming, the discovery of penicillin was both a triumph and a humbling experience. He was known for his modesty, often saying that he hadn't really "invented" penicillin; rather, nature did, and he had simply noticed it. But that's precisely what makes his story so inspiring. Fleming's discovery wasn't about flashy experiments or high-tech equipment—it was about paying attention, being curious, and not dismissing what others might consider an accident. It's a reminder that sometimes, great breakthroughs come from the most unexpected places, and they're often a matter of being open to seeing what's right in front of us.

The discovery of penicillin marked a turning point in human history. Before penicillin, there were no effective treatments for bacterial infections. People died from pneumonia, infected wounds, and diseases that we now consider easily treatable. Fleming's breakthrough gave the world hope and laid the foundation for the antibiotics that would follow, saving millions of lives and changing the way doctors approached infectious diseases.

It's easy to picture Fleming in that moment—standing over a petri dish, peering through his microscope, and realizing that the mold he was looking at was something special. He didn't know it then, but he was holding a key that would unlock a new era in medicine. His discovery wasn't just a scientific achievement; it was a gift to humanity, born out of an inquisitive mind, an eye for detail, and a little bit of luck. Today, when we take an antibiotic for an infection, we owe a debt of gratitude to that moment in Fleming's lab—a reminder of how a single observation can change the world.

The Development and Impact on Medicine

The story of penicillin didn't end with Alexander Fleming's accidental discovery in his cluttered lab. In many ways, it was just the beginning. Fleming had stumbled upon something incredible, but the path from that moment to the mass production of penicillin—a drug that would change the face of medicine—was long and full of challenges. It took a team of dedicated scientists, a world on the brink of war, and a fair share of human ingenuity to turn penicillin from a laboratory curiosity into a life-saving medicine.

After Fleming's discovery in 1928, the real challenge lay in transforming penicillin from a delicate mold into a reliable treatment. Fleming had shown that penicillin could kill bacteria, but he struggled to purify it and produce it in large quantities. That's where Howard Florey, Ernst Boris Chain, and Norman Heatley came in. It was the late 1930s, and Europe was on the edge of World War II. Amidst the uncertainty, Florey and his team at Oxford University saw potential in Fleming's work and decided to take on the daunting task of developing penicillin into a usable drug.

It wasn't easy. Imagine trying to extract enough penicillin from a batch of mold to treat even one person—it was a painstaking process. The team worked day and night, using every trick they could think of. They even enlisted the help of a local brewery, borrowing equipment to try and grow larger amounts of the mold. At one point, they were so desperate for penicillin that they collected the patients' urine to extract whatever penicillin was left and reuse it. It was far from glamorous, but their dedication paid off.

In 1941, they treated their first patient, a policeman who was suffering from a severe bacterial infection. The man's condition began to improve dramatically once he received penicillin, but, tragically, they didn't have enough of the drug to fully cure him. He ultimately succumbed to the infection. However, this early success was enough to prove that penicillin had the potential to save lives on an unprecedented scale. The challenge now was to produce it in large enough quantities.

The turning point came when the United States got involved. With the war raging, the U.S. government saw the strategic importance of penicillin—it could save soldiers' lives and get them back on the battlefield. American pharmaceutical companies were brought in, and with the support of the government, they began working on ways to mass-produce penicillin. By D-Day in 1944, penicillin was available in large quantities, and thousands of soldiers received it to treat their wounds and infections. It was often called "the wonder drug," and for good reason. What had once been a death sentence—a deep wound, pneumonia, or a staph infection—was now treatable with a few doses of this miraculous mold-derived substance.

The impact of penicillin on medicine was profound. Before its development, even minor infections could turn deadly. A simple scratch could lead to sepsis, childbirth carried significant risks of infection, and diseases like pneumonia and syphilis were often untreatable. Penicillin changed all of that. For the first time, doctors had a reliable way to fight bacterial infections, and the number of lives saved was staggering.

One of the most touching aspects of penicillin's impact was how it brought hope where there had once been only despair. Imagine being a parent in the early 20th century, watching your child suffer from a bacterial infection and knowing there was nothing you could do. Then, suddenly, there was a medicine that could stop the infection in its tracks. Penicillin wasn't just a medical breakthrough—it was a lifeline, a miracle that brought relief to millions.

The success of penicillin also sparked the golden age of antibiotics. Scientists were inspired by Fleming's mold and began searching for other substances that could fight bacteria. This led to the development of many more antibiotics, each targeting different types of infections. The ability to control bacterial infections revolutionized not only medicine but also surgery. Procedures that had once been too risky due to the threat of infection now became routine. The entire field of medicine was transformed, and life expectancy rose significantly as a result.

However, with great power came challenges. Over time, bacteria began to adapt, and antibiotic resistance emerged—a reminder that nature is always evolving. While penicillin had been a miracle, it wasn't invincible. This ongoing battle between medicine and bacteria is a testament to the complexity of the natural world and the need for continued innovation.

The development and impact of penicillin is a story of persistence, ingenuity, and the human spirit's refusal to accept defeat. It's about people like Alexander Fleming, who saw potential in a moldy petri dish, and scientists like Florey, Chain, and Heatley, who worked tirelessly to bring that potential to the world. It's about the countless lives saved, the families kept whole, and the hope restored. Penicillin didn't just change medicine—it changed humanity's relationship with disease, giving us the tools to fight back and win.

Why Penicillin Matters Today

Penicillin might seem like an old story, something from the past that only belongs in history books. But the truth is, penicillin still matters today—maybe even more than ever. It's not just about the discovery that changed the course of medicine; it's also about the ongoing fight to keep the benefits of this life-saving antibiotic alive in the face of new challenges. Penicillin's story is one that continues to evolve, reminding us of the power and limits of medical science.

Think about it for a moment. Before penicillin, something as simple as a minor scrape could become a death sentence if it got infected. It's easy to take for granted that we live in a world where bacterial infections are, for the most part, manageable. That confidence, that ability to treat an infection with a quick visit to the doctor, all started with penicillin. It was the first of its kind—a miracle drug that opened the door to modern antibiotics, changing how we think about disease and giving people hope where there had been none.

Even now, penicillin remains a cornerstone of medical treatment. It's still used to treat a wide range of bacterial infections, from strep throat to syphilis. Its impact is felt across the world, particularly in places where access to healthcare is limited and penicillin remains an affordable and effective treatment option. Penicillin has saved millions of lives, and it continues to save lives today.

But perhaps the most important reason why penicillin matters today is because of the lessons it teaches us about antibiotic resistance. When penicillin was first introduced, it worked like magic. Infections that had once been fatal suddenly became curable. But bacteria are clever, and over time, some of them started to find ways to survive even in the presence of penicillin. This is what we call antibiotic resistance—a growing problem that poses a serious threat to global health. The story of penicillin is also a cautionary tale, reminding us that we can't take antibiotics for granted.

I remember talking to my grandfather once about his childhood. He grew up in the 1930s, and he told me about a friend of his who died from a simple cut that got infected. He said that the doctors had done everything they could, but there was just no way to stop the infection. My grandfather always got emotional when he talked about that friend. It's hard to imagine that kind of helplessness today, but antibiotic resistance could bring us back to a time when we didn't have the tools to fight off infections. Penicillin matters because it symbolizes what we stand to lose if we're not careful—if we overuse antibiotics or fail to develop new ones.

On the flip side, penicillin also represents hope and human ingenuity. It reminds us that even when things seem grim, there's always a chance for a breakthrough. The discovery of penicillin was a moment of brilliance, and its development was a triumph of persistence and creativity. We need that same spirit today as we face the challenge of antibiotic resistance and look for new solutions.

Researchers are exploring new ways to combat resistant bacteria, from developing new antibiotics to finding alternative treatments, like bacteriophages—viruses that specifically target bacteria.

Penicillin also holds an important place in public health. It taught us the importance of making life-saving medicines accessible to everyone. During World War II, the mass production of penicillin became a priority, and it was made available to soldiers who needed it. After the war, efforts were made to ensure that civilians had access to it as well. This set the stage for the idea that healthcare should be a right, not a privilege, and it's a principle we're still fighting for today.

Why does penicillin matter today? Because it's a reminder of how far we've come and how much we still need to do. It's a testament to human curiosity, to the power of science, and to the importance of vigilance in the face of new challenges. When you look at that little pill, you're not just seeing an antibiotic—you're seeing a legacy. A story of people who refused to accept the limits of their time and who gave the world a gift that's still saving lives.

Penicillin matters because it's about more than just treating infections. It's about the human spirit—about our endless drive to find answers, to protect one another, and to turn even the smallest discoveries into something that can change the world. And as we continue to face new challenges in medicine, it's that same spirit that will carry us forward.

Chapter 1

How Penicillin Works

When Alexander Fleming discovered penicillin in 1928, he had no idea just how profoundly it would change the world. It was the key to unlocking a powerful weapon against bacterial infections, but how does it actually work? Understanding how penicillin fights bacteria can help us appreciate why this simple mold became such a game-changer in medicine.

Imagine bacteria as tiny houses surrounded by sturdy walls. These cell walls are essential for the bacteria's survival—they keep everything inside and protect them from the outside environment. Penicillin works by attacking these walls, essentially breaking them down so the bacteria can no longer survive. It's a bit like poking holes in a dam; eventually, the water will burst through and flood everything. Without a solid wall, bacteria simply can't hold themselves together, and they die.

The science behind this might sound complicated, but at its core, penicillin prevents bacteria from creating the molecules they need to build their walls. Bacteria have an enzyme that helps glue the bricks of their walls together, and penicillin cleverly blocks this enzyme. Without their "bricklayer," the bacteria are unable to finish building their walls, and they collapse. It's incredibly effective—and it's this simplicity that made penicillin such a powerful tool in the fight against infections.

I remember reading about a young boy named Johnny, who had a serious bacterial infection before penicillin was available. Johnny was only ten years old, and his family had tried everything they could, but his condition kept getting worse. It was heartbreaking for his parents—they were watching their child fade away, and there was nothing they could do. When penicillin became available, everything changed. Johnny was one of the first patients to receive it, and within days, he started to recover. His fever dropped, his strength returned, and soon enough, he was back to playing in the yard. This transformation seemed almost miraculous, but it all came down to the simple way penicillin attacked the bacteria causing his infection.

What makes penicillin even more remarkable is how specific it is. It targets bacteria, but it doesn't harm human cells. That's because our cells are built differently; we don't have the same kind of cell walls that bacteria do. So while penicillin is busy tearing down the bacterial walls, it leaves our cells completely untouched. It's like having a smart missile that only goes after the enemy, sparing everything else. This precision is one of the reasons penicillin was such a breakthrough—it could wipe out infections without causing serious harm to the patient.

There's also something poetic about the fact that penicillin, derived from a humble mold, turned out to be a perfect adversary for bacteria. Nature provided both the problem and the solution. The mold that produced penicillin was fighting its own battle against bacteria, trying to carve out a little space to grow. In a way, the mold's survival tactic became humanity's best weapon against some of the most dangerous infections.

But it's important to remember that while penicillin is powerful, it has its limits. It only works against bacteria, not viruses. So while it can cure a strep throat, it won't help with the common cold or the flu. I've heard countless stories of people taking antibiotics for viral infections, not realizing that they're completely ineffective. This misunderstanding is part of why antibiotic resistance is such a growing problem today.

When we look at how penicillin works, we're seeing a beautiful dance of biology—a small molecule produced by a mold that can dismantle the defenses of bacteria, saving lives in the process. It's simple, elegant, and profoundly effective. Penicillin's story is a reminder of how sometimes the most extraordinary solutions come from the simplest places, and how even a bit of mold can hold the power to change the world.

1.1 The Science Behind Penicillin

To truly appreciate the magic of penicillin, it's helpful to take a closer look at the science behind it. Imagine a time when there were no antibiotics—a time when even a small cut could lead to a life-threatening infection, and doctors were often powerless in the face of bacterial diseases. Then, almost like a miracle, penicillin came along, armed with a natural ability to fight off harmful bacteria. But how does this seemingly simple substance work its wonders?

At its core, penicillin is a type of molecule known as an antibiotic, which means it can kill or stop the growth of bacteria. The story of penicillin begins with mold—the kind you might see on an old loaf of bread. Alexander Fleming, who discovered penicillin, noticed something curious: the mold seemed to be keeping bacteria at bay. He decided to investigate further, and what he found was groundbreaking. The mold, called *Penicillium notatum*, was producing a chemical that prevents bacteria from growing around it. This chemical was penicillin, and it turned out to be a natural enemy of many dangerous bacteria.

But let's dig a little deeper into how penicillin works at the microscopic level. Bacteria, unlike human cells, are surrounded by a rigid cell wall that helps protect them and keep their shape. Think of this cell wall like the outer shell of a castle, keeping the bad stuff out and holding everything together. The walls are made up of molecules linked together like bricks in a fortress, and bacteria have special enzymes that act as the builders, assembling these bricks into a solid structure. Penicillin's brilliance lies in its ability to interfere with these builders.

Penicillin binds to the enzymes that bacteria use to build their cell walls, effectively jamming up the works. It's like putting glue in the lock of a door—suddenly, the key (in this case, the enzyme) can't turn, and the door stays open. Without their enzymes working properly, bacteria can't complete their cell walls, and without those walls, they become vulnerable. Eventually, they burst and die because they can't contain themselves anymore. It's a little like inflating a balloon until it pops—without the wall, the bacteria just can't hold on.

One fascinating aspect of penicillin's action is how specific it is. The molecule is structured in such a way that it fits perfectly into the active site of the bacterial enzyme—almost like a key fitting into a lock. This precision means that penicillin can specifically target bacteria without harming our own cells, which lack the same type of cell walls. It's a targeted strike, one that goes after the bad guys without causing collateral damage.

I remember learning about this in biology class, and my professor told us a story that stuck with me. She said that before penicillin was discovered, surgeons would sometimes refuse to operate on even simple abscesses because they knew that once the infection spread, they would have no way of stopping it. Penicillin changed that. It gave doctors the power to fight back, to save lives that would have otherwise been lost. It's hard to imagine the kind of relief and joy doctors must have felt when they finally had a tool to combat infections that had plagued humanity for centuries.

What's even more amazing is that penicillin's method of attack inspired the development of other antibiotics. Scientists realized that if one mold could produce a substance capable of stopping bacterial growth, perhaps there were others. They began to search for new antibiotics, leading to the discovery of many more life-saving drugs, each targeting bacteria in a unique way.

Penicillin's success wasn't just about its ability to kill bacteria; it was also a story of human ingenuity. When Fleming made his discovery, penicillin was difficult to produce in large quantities. It took years of research and collaboration, particularly during World War II, to figure out how to mass-produce it. Scientists worked tirelessly to scale up production, finding better strains of mold and optimizing the conditions for growth. Their efforts paid off, and by the end of the war, penicillin was available to treat wounded soldiers and civilians alike.

The science behind penicillin is a story of nature's brilliance and humanity's determination. It's a reminder of how something as small as a mold spore can hold the key to saving millions of lives, and how a curious mind can take that key and unlock a world of possibilities. When we understand how penicillin works, we not only learn about the biology of bacteria and antibiotics, but we also gain insight into the incredible journey from discovery to life-saving medicine. It's a journey that started in a messy lab with a forgotten petri dish and ended up changing the world.

1.2 Mechanism of Action: How It Kills Bacteria

When we think about antibiotics, it's easy to imagine them as warriors in a battle—fighting off the enemy bacteria that make us sick. But how exactly does penicillin go about its mission? The science behind its mechanism of action is fascinating, and it's all about undermining the enemy's most critical defense: its cell wall.

To understand penicillin's power, imagine bacteria as tiny fortresses. These fortresses are protected by a sturdy wall, which keeps them safe and holds them together. Without this wall, the bacteria would fall apart, much like a castle without its outer defenses. Penicillin's magic lies in how it breaks down this essential barrier, effectively turning those fortified bacteria into vulnerable targets.

The process starts with penicillin sneaking into the bacteria's building site—the enzymes that construct the cell wall. These enzymes are like skilled masons, laying down bricks (which are actually molecules called peptidoglycans) to form a solid structure. What penicillin does is brilliant in its simplicity: it binds to these enzymes and blocks their ability to do their job. It's as if someone walked into a construction site and took away all the workers' tools, leaving the wall half-built and collapsing under pressure.

I remember a story that a microbiologist once told during a seminar. He compared penicillin to a wrench thrown into the gears of a machine. The bacterial cell wall, he explained, is a well-oiled machine, operating smoothly as long as all the parts are in place. When penicillin binds to the enzyme, it acts like that wrench, jamming the whole mechanism and bringing the wall-building to a grinding halt. Without a fully formed wall, the bacteria can't contain its internal contents, and eventually, it bursts—killed by its own inability to maintain structure.

The elegance of penicillin's mechanism also lies in how selective it is. It targets an enzyme called transpeptidase, which is essential for bacteria but doesn't exist in human cells. This means that while penicillin is wreaking havoc on the bacteria's ability to build a wall, it leaves human cells completely unscathed. It's a bit like a perfectly targeted weapon that knows exactly which structures to destroy while leaving everything else intact—a precision that makes penicillin a true marvel of nature.

One of the most fascinating parts of this story is how penicillin's action was discovered through pure serendipity. When Alexander Fleming noticed that bacteria couldn't grow around a certain mold, he didn't know exactly why. It wasn't until later that scientists unraveled the detailed mechanism—how penicillin blocks the enzymes responsible for creating the bacterial cell wall. Understanding this process took years of careful experimentation, and it was a triumph of curiosity and persistence.

The human stories behind penicillin's action are just as compelling as the science. During World War II, the mass production of penicillin saved countless lives. Soldiers with infected wounds could suddenly be treated, and people who might have died from simple infections recovered quickly. The mechanism by which penicillin kills bacteria—by tearing down their defenses—was a key reason it became known as a miracle drug.

However, penicillin's success also led to some unexpected challenges. Bacteria, like all living things, adapt. Some began to evolve defenses against penicillin, producing enzymes called beta-lactamases that could break down the antibiotic before it could do its job. It's a reminder that while penicillin was incredibly effective, the battle against bacterial infections is an ongoing one. The bacteria fight back, and we have to keep finding new ways to outsmart them.

Penicillin's mechanism of action shows us the power of nature and science working together. The mold that produces penicillin evolved this ability as a means of survival—it was fighting off bacteria long before humans discovered its secret. By studying and understanding this natural weapon, we found a way to harness it for our own battles against infection. It's a story of discovery, innovation, and the relentless pursuit of knowledge—a story that began with a petri dish and a stroke of luck, but ended up transforming modern medicine forever.

1.3 Types of Bacteria Penicillin Targets

When you hear about penicillin, it's easy to picture it as a kind of superhero—rushing in to defeat harmful bacteria and save the day. But like any good superhero, penicillin has its specific foes. Not all bacteria are vulnerable to its powers; penicillin works best against certain types, mainly those with a particular kind of structure. Let's take a look at the enemies penicillin targets, and why it works so effectively against them.

To start, we need to understand that bacteria come in two main groups: Gram-positive and Gram-negative. These classifications, named after the Danish scientist Hans Christian Gram, relate to the structure of the bacterium's cell wall. The cell wall is like a protective armor that determines how the bacteria look under a microscope when stained with a special dye. Gram-positive bacteria have thick cell walls that are easy to see, while Gram-negative bacteria have a thinner wall hidden behind an extra outer membrane.

This difference makes a big impact on how penicillin works.

Penicillin is particularly effective against Gram-positive bacteria, which include some of the most notorious culprits behind common infections. Think of *Streptococcus* bacteria, which are responsible for illnesses like strep throat, scarlet fever, and some types of pneumonia. These bacteria have thick cell walls made of peptidoglycan, which penicillin can easily target and destroy. Because Gram-positive bacteria lack the extra outer membrane that Gram-negative bacteria have, penicillin can reach their vulnerable cell wall-building enzymes more easily, disrupting the construction of their protective armor.

I remember learning about this in college, and our professor painted a vivid picture: he said to think of Gram-positive bacteria as a medieval castle with a single thick wall. Penicillin acts like a siege weapon, breaking down the wall until the entire castle crumbles. On the other hand, Gram-negative bacteria are more like a modern fortress, complete with multiple layers of defense, making it much harder for penicillin to breach.

One of the most famous Gram-positive bacteria that penicillin targets is *Staphylococcus aureus*, which can cause a range of infections from minor skin problems to more serious conditions like pneumonia and sepsis. In the early days of penicillin's use, it was a game-changer in treating staph infections, which were often life-threatening before antibiotics became available. Imagine being a doctor in the 1940s, finally having a tool that could stop these terrible infections. It must have felt like a miracle—patients who were once on the brink of death began to recover within days of receiving penicillin.

However, not all bacteria are easily conquered. Gram-negative bacteria, like *Escherichia coli* (E. coli) or *Pseudomonas aeruginosa*, have that additional outer membrane, which acts like a shield to prevent penicillin from getting inside. It's almost as if they've evolved an extra layer of armor specifically to keep antibiotics out. Because of this, penicillin has a much harder time targeting Gram-negative bacteria. This is one reason why researchers have developed different types of antibiotics over the years—ones that can break through these extra defenses.

There's also a group of bacteria called anaerobes, which live in environments without oxygen. Some of these bacteria can be targeted by penicillin as well, particularly those that cause infections in wounds or deep within tissues. For instance, penicillin has been used to treat *Clostridium perfringens*, a bacterium responsible for gas gangrene—a severe and potentially deadly infection often seen in soldiers during wartime before antibiotics were widely available.

One interesting story about penicillin's role in targeting specific bacteria comes from World War II. Soldiers with wounds often developed deadly infections, and there were few effective treatments available. Penicillin's ability to target *Streptococcus* and *Clostridium* bacteria made it invaluable on the battlefield. Doctors were finally able to treat wound infections effectively, and countless lives were saved as a result. It's amazing to think how a simple mold, discovered almost by accident, could make such a difference in such dire situations.

Penicillin is not a cure-all, but its effectiveness against Gram-positive bacteria and certain anaerobes has made it a cornerstone of modern medicine. Its discovery and development opened the door to a whole new world of antibiotics, each with its own targets and mechanisms of action. The story of penicillin is not just about the bacteria it fights but also about the lives it has saved, and the countless people who were given a second chance because of this powerful drug.

Today, while penicillin remains a critical tool in fighting bacterial infections, the rise of antibiotic resistance reminds us that the battle is ongoing. Some bacteria have evolved ways to resist penicillin, which is why it's so important to use antibiotics responsibly. Still, understanding the types of bacteria penicillin targets helps us appreciate its role as one of the first true heroes of modern medicine—turning the tide in our fight against infectious diseases and laying the foundation for the development of more advanced antibiotics in the future.

1.4 Penicillin vs. Other Antibiotics

When it comes to antibiotics, penicillin often feels like the household name—it's the one many of us heard about first, the original hero of medicine that changed the world. But what about all the other antibiotics that followed? How does penicillin stack up against them, and why do doctors sometimes reach for other options?

To understand this, let's think of penicillin as a pioneer. It was the first antibiotic discovered, opening doors to a new era of medicine. It's like the Model T of antibiotics—revolutionary and reliable, but as time went on, newer and more specialized antibiotics came along, each with its own unique capabilities. Today, the world of antibiotics is vast, each designed to target different types of infections or bacteria with varying degrees of success.

Penicillin is part of a family of antibiotics known as beta-lactams. This group includes other antibiotics like amoxicillin and cephalosporins, all of which share a similar structure and mechanism: they work by breaking down the bacterial cell wall. Penicillin was groundbreaking because it was the first to do this, but other antibiotics were developed to overcome some of the challenges that penicillin faced. For example, amoxicillin is like penicillin's more modern cousin—it's broader in its spectrum and is often used when doctors need something a bit more versatile.

I remember a story my grandmother told me about her childhood. She had scarlet fever in the late 1940s, and penicillin was the miracle that saved her. Back then, it was one of the few tools doctors had to fight dangerous bacterial infections. Nowadays, we have many more options. Take macrolides, for instance—this class includes antibiotics like erythromycin and azithromycin. They work differently from penicillin; instead of targeting the cell wall, they stop bacteria from making the proteins they need to grow. This makes them useful for bacteria that penicillin can't touch, particularly those without traditional cell walls.

Then there are tetracyclines, such as doxycycline. These antibiotics are especially effective against a wide variety of infections, including some that penicillin isn't as good at treating, like Lyme disease or certain respiratory infections. If penicillin is the trusty sword in the battle against bacteria, tetracyclines are more like a Swiss army knife—versatile and ready to tackle a range of challenges.

One of the biggest differences between penicillin and other antibiotics is how they deal with resistance. Bacteria are incredibly clever, and over time, some of them developed ways to resist penicillin. They produce an enzyme called beta-lactamase that breaks down the antibiotic before it can work. Imagine penicillin as a key trying to unlock a door, and the bacteria have figured out a way to change the lock. In response, scientists developed new antibiotics like methicillin, which were designed to be resistant to this enzyme. Unfortunately, even this wasn't foolproof—bacteria like *Staphylococcus aureus* eventually evolved into MRSA (methicillin-resistant Staphylococcus aureus), a superbug that's tough to treat.

Another important group to mention is fluoroquinolones, such as ciprofloxacin. These antibiotics work by interfering with the bacteria's DNA, stopping them from replicating. They're particularly effective against Gram-negative bacteria—those with the extra outer membrane that makes them hard for penicillin to penetrate. So, when someone has a urinary tract infection caused by E. coli, a doctor might choose ciprofloxacin over penicillin because it's more effective for that type of bacteria.

It's fascinating how each antibiotic has its own strengths and weaknesses, and the choice of which one to use often comes down to the specific type of bacteria causing an infection. If you have a simple strep throat, penicillin might be the best and simplest option. But if you're dealing with a more complex infection or one caused by a resistant bacterium, the doctor might choose something from a different class.

The development of antibiotics beyond penicillin was not just about fighting different bacteria, but also about dealing with the challenges of resistance. In the 1950s and 60s, as penicillin-resistant bacteria became more common, scientists had to innovate. This led to the creation of entirely new classes of antibiotics, each with different mechanisms of action. It's a bit like an arms race—bacteria evolve, and we develop new antibiotics to keep up.

But as powerful as these other antibiotics are, they all owe a debt to penicillin. Without Alexander Fleming's accidental discovery in 1928, we might never have unlocked the potential of antibiotics at all. Penicillin was the starting point—the trailblazer that showed us it was possible to fight back against bacterial infections. It taught us that we could harness nature to cure disease, and that knowledge paved the way for everything that followed.

Today, doctors have a whole arsenal of antibiotics to choose from, and the choice depends on many factors: the type of bacteria, the site of infection, the patient's history, and the potential for resistance. Penicillin may no longer be the only weapon in our fight against bacterial infections, but it remains one of the most important. Its story is a reminder of the power of scientific discovery and the ongoing battle against the ever-adapting world of bacteria.

Chapter 2

Medical Uses of Penicillin

When we think about penicillin, it's easy to see it as just a pill or an injection that helps us fight off infections. But its medical uses extend far beyond that simple view. Penicillin has played a monumental role in transforming medicine and has saved countless lives since its discovery. Let's take a journey through the various ways this antibiotic has been utilized in the medical field, making it a cornerstone of modern healthcare.

Imagine the scene in the 1940s when penicillin first became widely available. World War II was raging, and soldiers were suffering from wounds that often turned deadly due to bacterial infections. Before penicillin, these infections frequently led to amputations or death. But as penicillin became accessible, it started to change the battlefield and the home front alike. I remember my grandfather telling me how, after the war, he felt like he had been given a second chance. Many of his fellow soldiers who survived injuries credited penicillin with saving their lives.

The magic of penicillin lies in its ability to target and kill bacteria, which makes it particularly effective against certain infections. It's used primarily to treat infections caused by Gram-positive bacteria, such as streptococci and staphylococci. Conditions like strep throat, skin infections, and even pneumonia have found effective treatments through this antibiotic. Penicillin is like a seasoned warrior—familiar and skilled in battling these familiar foes.

Let's take a closer look at one of the most common uses of penicillin: treating strep throat. Picture a parent bringing a child to the doctor's office with a sore throat and fever. After a quick examination, the doctor suspects strep throat and orders a rapid strep test. If it comes back positive, penicillin is often the first choice for treatment. It's effective, quick, and usually well-tolerated. The child can be back to their usual, rambunctious self in no time, thanks to this miracle drug.

Penicillin also plays a critical role in treating more severe infections. For instance, in cases of bacterial meningitis, a serious inflammation of the protective membranes covering the brain and spinal cord, penicillin can be a lifesaver. In my community, I remember hearing about a college student who was diagnosed with bacterial meningitis. The swift administration of penicillin helped prevent a life-threatening situation. Stories like this remind us just how vital penicillin is in urgent medical settings.

Another area where penicillin shines is in the treatment of syphilis, a sexually transmitted infection that can lead to severe health complications if left untreated. Penicillin has been the go-to treatment for syphilis for decades, and its effectiveness in curing the infection has made it a key player in public health efforts. This was particularly evident in the early 20th century when syphilis was a major health crisis. Once penicillin became widely available, it led to a significant decline in syphilis rates, helping to restore health to many individuals and families.

But it's not just about treating established infections; penicillin has also proven useful in preventative medicine. For example, during certain surgeries, doctors may prescribe penicillin as a prophylactic measure to prevent potential infections. I recall a friend who had knee surgery and was given penicillin beforehand. Thanks to this precaution, she was able to recover without any postoperative infections, which can sometimes complicate the healing process.

While penicillin is a powerhouse in the world of antibiotics, it's important to note that it doesn't work for every type of infection. It's ineffective against viral infections like the common cold or flu. Understanding the right context for penicillin's use is crucial for both healthcare providers and patients. A personal anecdote comes to mind of a time when I rushed to the doctor with what I thought was a bad throat infection. The doctor explained that it was viral, and prescribing penicillin wouldn't help me. This was a valuable lesson on the importance of targeted treatment—penicillin is not a one-size-fits-all solution.

As penicillin became more widely used, the medical community also learned more about antibiotic resistance. Some bacteria evolved to resist the effects of penicillin, which led to the development of new antibiotics and treatment strategies. Yet, despite these challenges, penicillin remains a staple in treating bacterial infections, thanks to its proven effectiveness and relatively low cost.

Today, the uses of penicillin continue to expand. Ongoing research explores its potential in treating new types of infections and even its role in fighting certain cancers. The adaptability of penicillin reminds us that science is a continually evolving field, with discoveries that can lead to novel applications for existing medications.

Reflecting on penicillin's journey in medicine is a testament to the power of innovation and discovery. From its humble beginnings in Alexander Fleming's laboratory to its life-saving applications in modern healthcare, penicillin has been a beacon of hope in our fight against bacterial infections. Each prescription and each success story is a reminder of its impact—a drug that not only changed the course of medical history but continues to play a vital role in ensuring our health today.

2.1 Common Infections Treated by Penicillin

When we think about antibiotics, penicillin often comes to mind as a trusted ally in the battle against infections. Its discovery revolutionized medicine, providing us with the means to treat various bacterial infections that once posed serious health risks. Let's explore some of the common infections treated by penicillin and the impact it has had on countless lives.

One of the most familiar infections that penicillin effectively combats is strep throat. It's a common condition, especially among children. Picture a young child, complaining of a sore throat, with a fever and that distinctive grumpy demeanor. Parents often find themselves at the doctor's office, seeking answers. When the diagnosis is confirmed—strep throat—the doctor usually prescribes penicillin. Within just a few days, the child starts to feel better, able to return to school and play, thanks to this simple yet powerful antibiotic.

Another infection that penicillin tackles head-on is pneumonia, a serious lung infection that can be particularly dangerous for young children and the elderly. I recall a family friend who was hospitalized due to pneumonia. It was a scary time, as her health deteriorated rapidly. However, once she was treated with penicillin, she began to recover. Watching her regain her strength reminded me of the antibiotic's remarkable ability to fight such life-threatening infections.

Skin infections are also common culprits that penicillin can treat effectively. Infections like cellulitis, which affects the deeper layers of the skin, can occur due to cuts or insect bites. These infections can lead to swelling, redness, and pain, and without treatment, they can worsen significantly. A neighbor of mine once experienced a cellulitis infection after getting a scratch while gardening. Thanks to penicillin, the infection was brought under control, allowing her to return to her beloved garden in no time.

Penicillin has proven particularly effective in treating syphilis, a sexually transmitted infection that has been a concern for centuries. The history of syphilis is quite fascinating; before penicillin became widely available, treatments were often rudimentary and sometimes harmful. With penicillin, patients have a reliable cure. When someone is diagnosed with syphilis today, penicillin can clear the infection swiftly and safely, offering a fresh start. It's hard not to appreciate how far we've come in managing such infections.

Ear infections, particularly otitis media, are also commonly treated with penicillin. Many parents can relate to the sleepless nights caused by a child tugging at their ear in pain. A simple visit to the doctor often results in a prescription for penicillin, bringing relief within days. I remember when my niece had a nasty ear infection. After just one day on the antibiotic, her mood improved dramatically. The joy of seeing her playful and free of discomfort was a reminder of the magic of penicillin in treating everyday ailments.

In addition to these infections, penicillin is also used to prevent infections in certain circumstances. For example, during surgeries, it may be administered prophylactically to ensure that any potential bacterial invaders are kept at bay. This preventative approach has saved many patients from post-surgical complications, illustrating how penicillin serves not just as a treatment but also as a shield.

However, while penicillin is an effective tool against many infections, it's essential to remember that it isn't a cure-all. It works wonders for specific bacterial infections, but it won't help with viral infections like the flu or the common cold. I recall one time when I rushed to the doctor, convinced I had a bacterial infection. After a quick examination, the doctor explained that my symptoms were due to a virus. While I was disappointed I wouldn't be getting a prescription for penicillin, I learned a valuable lesson about the importance of appropriate treatment.

The continued effectiveness of penicillin reminds us of the importance of using antibiotics responsibly. With the rise of antibiotic resistance, understanding when to use penicillin is crucial. Healthcare providers play a key role in educating patients about the right use of antibiotics to ensure they remain effective for generations to come.

As we delve into the world of penicillin, we find it's not just a medication but a symbol of hope. It represents the advancements in medical science that allow us to treat common infections effectively and improve our quality of life. Each successful treatment story adds to the legacy of penicillin, making it a true hero in the realm of medicine. Whether it's a child recovering from strep throat or a patient overcoming pneumonia, the impact of penicillin resonates through countless lives, reminding us of its vital role in healthcare today.

2.2 Different Types of Penicillin

When we think of penicillin, we often imagine a single miracle drug that has saved countless lives since its discovery. However, the reality is much richer and more complex. There are several different types of penicillin, each tailored for specific uses and bacteria. Let's dive into the various types of penicillin and explore their unique characteristics, with some stories that bring these medicines to life.

First up is Penicillin G, often considered the gold standard in the penicillin family. Developed soon after the initial discovery, Penicillin G is administered intravenously or intramuscularly, making it incredibly effective for treating serious infections. I remember speaking with a doctor who recounted a particularly challenging case of endocarditis, an infection of the heart's inner lining. The patient was gravely ill, and after starting a course of Penicillin G, the transformation was remarkable. Within days, the patient's fever broke, and they began to regain their strength. It was a reminder of how powerful this form of penicillin can be in dire situations.

On the other hand, we have Penicillin V, which is taken orally. This version is a more stable compound, allowing it to be absorbed better in the digestive system. It's commonly prescribed for less severe infections, like strep throat or mild skin infections. I recall my sister's experience with strep throat as a child. After being diagnosed, she was given Penicillin V in liquid form. I still remember her joy when the doctor told her she could take the medication at home instead of getting an injection. Watching her recover within a few days was reassuring for our family, knowing that such a simple solution could make such a difference.

Another notable member of the penicillin family is Ampicillin. This type has a broader spectrum of activity compared to Penicillin G, meaning it can tackle a wider variety of bacteria. Ampicillin is often used in cases where the infection is more complex, such as respiratory infections or urinary tract infections. A family friend who had recurrent urinary infections swore by Ampicillin. It became her go-to when other treatments had failed, showcasing how personalized medical treatment can be.

Next on the list is Amoxicillin, which is perhaps one of the most widely recognized antibiotics today. It's a cousin of Ampicillin, with a similar broad spectrum of action but with better absorption when taken orally. My neighbor's son once had a particularly nasty ear infection, and after trying several treatments, the pediatrician prescribed Amoxicillin. The transformation was quick; he was back to his playful self in no time. Amoxicillin is often the first line of defense against common infections in children, and many parents have stories of how it helped their little ones bounce back.

In the world of penicillins, we also find some less commonly discussed types, like Oxacillin and Cloxacillin. These are resistant to penicillinase, an enzyme some bacteria produce to defend themselves against antibiotics. This makes them effective against certain strains of bacteria that have developed resistance. A colleague shared a story about a friend who faced a stubborn staphylococcal infection. After various unsuccessful treatments, the doctors turned to Oxacillin, which ultimately cleared the infection. It was a testament to how diverse the penicillin family is in addressing different bacterial challenges.

While we often hear about these specific types of penicillin, it's important to remember the context in which they are used. Different infections require different approaches, and healthcare providers must consider a variety of factors when deciding on the best treatment. Each penicillin type has its own advantages and is suited for specific situations, allowing for tailored care that can lead to the best possible outcomes.

As we explore the landscape of penicillin, it's evident that these antibiotics are more than just a one-size-fits-all solution. They are a collection of specialized tools in a physician's arsenal, each with its own unique story. From Penicillin G's heroic efforts in critical cases to Amoxicillin's gentle touch for children's infections, these different types of penicillin illustrate the remarkable diversity and effectiveness of antibiotic treatments. It's a reminder that, in medicine, one size rarely fits all, and the right medication can make all the difference in a patient's journey to recovery.

2.3 Dosage Guidelines and Administration

When it comes to antibiotics like penicillin, understanding the right dosage and administration is crucial. It's not just about taking a pill; it's about ensuring the body receives the right amount at the right time to effectively combat an infection. Let's explore the nuances of dosage guidelines and administration, using relatable stories to highlight their importance.

Imagine a busy mom named Sarah, whose son came home from school one day with a fever and a sore throat. After a quick trip to the doctor, she received a prescription for penicillin to treat his strep throat. The doctor explained the importance of following the dosage guidelines closely—taking the medication exactly as prescribed would ensure that the bacteria causing the infection would be effectively eliminated. This was no small detail; it meant the difference between a speedy recovery and a lingering illness.

For penicillin, the dosage often varies based on factors like age, weight, and the severity of the infection. For children, doctors typically calculate the dosage based on weight, usually in milligrams per kilogram. So, for Sarah's son, the doctor calculated the appropriate dose and instructed her to administer it three times a day for ten days. This structure wasn't just a random number; it was designed to keep the antibiotic levels steady in his bloodstream, ensuring the medication would effectively fight off the bacteria.

On the other hand, adults typically receive a fixed dose, depending on the type of infection. For example, a common recommendation for mild to moderate infections could be 250 to 500 mg every six hours. However, if the infection is more severe, the doctor might prescribe higher doses. It's interesting to note how the body's response can vary so much; what works for one person might need adjustment for another. My uncle had a persistent skin infection that required him to be on a higher dose of penicillin for a longer duration. The doctor monitored his progress, making adjustments to ensure he was on the right path to recovery.

The administration method also plays a critical role. While some forms of penicillin, like Penicillin V, are available in oral tablets or liquid forms, others, like Penicillin G, are administered by injection. This route is often necessary for severe infections, as it allows the medication to enter the bloodstream quickly and effectively. I remember visiting a friend in the hospital who was receiving Penicillin G through an IV. The nurses were meticulous about checking the dosage and the rate of administration, ensuring that everything was just right. It was impressive to see how critical these details are in a hospital setting, especially for patients with serious conditions.

Timing is another crucial aspect. Consistency in taking penicillin helps maintain effective drug levels in the body, minimizing the risk of bacterial resistance. Sarah made a point of setting reminders on her phone to ensure her son took his medicine at the same time each day. Her diligence paid off, as he started feeling better within days, allowing him to return to his usual energetic self.

One challenge that arises in the administration of antibiotics is the issue of missed doses. Life can be hectic, especially with children. If a dose is missed, it's essential to take it as soon as you can remember. However, if it's close to the time for the next dose, it's usually best to skip the missed dose and resume the regular schedule. It's easy to feel anxious about this, but the key is to follow the healthcare provider's instructions.

Ultimately, dosage guidelines and administration of penicillin embody a balance between science and personal experience. These guidelines are meticulously crafted based on years of research and clinical practice. Yet, behind every prescription, there are real people like Sarah and her son, navigating the challenges of illness and recovery. Whether it's a mother caring for her child or an adult managing their own health, understanding the importance of dosage and administration becomes a vital part of the healing journey.

As we reflect on these guidelines, it's clear that penicillin is more than just a medicine. It's a lifeline for many, and getting the details right is essential for effective treatment. Each dosage and each administration story underscores the human side of medicine, reminding us that health is a journey best taken with care, attention, and a little help from modern science.

2.4 Special Considerations: Children, Elderly, and Pregnant Women

When it comes to medication, especially something as impactful as penicillin, not everyone is treated the same. Factors like age and specific health conditions can greatly influence how penicillin is prescribed and administered. This isn't just a medical guideline; it's a matter of ensuring safety and effectiveness for everyone involved. Let's dive into some of the special considerations that healthcare providers keep in mind when prescribing penicillin to children, the elderly, and pregnant women.

Imagine a concerned mother, Lisa, who brings her young daughter, Mia, to the doctor with a severe ear infection. The doctor prescribes penicillin, but before anything is finalized, they discuss how children metabolize medications differently. For children, dosages are often calculated based on their weight. Lisa learned that the standard recommendation is about 20 to 50 mg of penicillin per kilogram of body weight for treating common infections. This means that Mia, weighing around 30 kg, would need a carefully calculated dose to ensure she receives just the right amount. It's a precise balancing act that emphasizes how seriously healthcare providers take these guidelines.

The doctor also talks to Lisa about potential side effects. Children might experience upset stomachs or rashes, which can sometimes occur with penicillin. It's important for parents to know what to watch for. Lisa felt reassured when the doctor explained how to monitor Mia and what to do if any adverse reactions appeared. This kind of conversation fosters trust and helps parents feel more involved in their children's care, a vital part of the healing process.

Now, let's shift our focus to the elderly. My grandmother, at 85, has her fair share of health issues, including high blood pressure and diabetes. When she developed a respiratory infection, her doctor opted for penicillin but took extra precautions. Older adults often have different physiological responses to medications. They might process drugs more slowly due to changes in liver and kidney function. So, while a younger adult might receive a standard dose, my grandmother's doctor prescribed a lower dosage to start, gradually adjusting based on her response to treatment. It's a thoughtful approach that highlights the importance of monitoring and adjusting medications as people age.

In my grandmother's case, the doctor also discussed potential interactions with her existing medications. It's a reality that many elderly patients are on multiple prescriptions, so ensuring that penicillin wouldn't negatively interact with her other drugs was essential.

This kind of careful consideration is vital in promoting safety and efficacy in treatment.

Pregnant women face unique challenges when it comes to medication, and penicillin is no exception. Sarah, a friend of mine, was pregnant and developed a bacterial infection. She was understandably worried about the safety of taking antibiotics. Her doctor reassured her that penicillin is generally considered safe during pregnancy, especially in treating infections that could pose a greater risk if left untreated. However, there are still important precautions to consider.

Sarah's doctor emphasized that not all antibiotics are safe, and choosing the right one is critical. They talked through her medical history and any allergies she had. It's a conversation that highlights the importance of personalized care, ensuring both the mother and her unborn child remain safe throughout the treatment process. This kind of dialogue makes a world of difference for expectant mothers navigating the complexities of health and pregnancy.

For each of these groups—children, the elderly, and pregnant women—the focus remains on individualized care. It's about recognizing that penicillin is a powerful tool, but one that must be used thoughtfully and carefully. Each case tells a story of trust and understanding, where healthcare providers work closely with patients and their families to ensure the best outcomes.

Ultimately, special considerations in prescribing penicillin underscore the need for a personalized approach to healthcare. Every patient comes with their unique circumstances, and adapting treatment plans accordingly is not just a practice; it's an art form. This is what makes medicine not only a science but also a deeply human endeavor. Whether it's a worried parent, a caring grandchild, or an expectant mother, everyone deserves attention and compassion in their healthcare journey.

Chapter 3

The Impact of Penicillin on Global Health

When we think about the advances in medicine over the past century, it's hard to overlook the monumental role penicillin has played. It's not just a story of a mold discovered by accident; it's a tale of how a single antibiotic reshaped the landscape of global health, saving countless lives and changing the way we view infectious diseases.

Imagine being a doctor in the early 20th century, faced with patients suffering from severe bacterial infections. The options were limited, and the outcomes were often grim. Then came the 1940s, and with it, penicillin burst onto the scene like a superhero swooping in to save the day. Suddenly, doctors had a powerful weapon against infections that once claimed lives with ruthless efficiency. It felt as if a heavy veil had been lifted from medicine, allowing practitioners to treat what were once considered fatal conditions.

Let's take a moment to reflect on a personal story that highlights this shift. A close family friend, John, shared his experience growing up in the 1940s, where childhood infections were a serious threat. He recounted the fear his parents felt when he contracted pneumonia. In the past, this could have been a death sentence. But because penicillin was available, he was treated successfully, allowing him to live a healthy life and raise a family of his own. John's story isn't just one of survival; it's a testament to how penicillin helped shift societal attitudes toward bacterial infections, enabling families to breathe easier knowing that effective treatments were at their disposal.

The impact of penicillin goes beyond individual stories. It sparked a revolution in the field of medicine. Hospitals began to stock antibiotics as a standard part of their treatment arsenal, leading to the establishment of protocols and guidelines for their use. This new approach dramatically reduced mortality rates from bacterial infections. Before penicillin, a simple cut could lead to a life-threatening infection. With penicillin, the fear of minor injuries diminished, transforming the way people viewed health and safety.

But penicillin's impact wasn't confined to developed nations. As it became widely available, it reached parts of the world that desperately needed it. The World Health Organization launched initiatives to distribute antibiotics to developing countries, transforming health outcomes on a global scale. I recall reading about a mission in the 1950s, where healthcare workers traveled to remote villages in Africa, providing access to penicillin for treating common infections. It was a game-changer for communities struggling with the burden of disease.

Yet, the journey wasn't without challenges. The rise of antibiotic resistance became a concerning backdrop to penicillin's success story. As penicillin became widely used, some bacteria developed resistance, leading to complications in treatment. This wasn't the end of the story, though. It sparked a new era of research and innovation in antibiotic development, pushing scientists to explore new avenues and refine existing treatments. It was a reminder that while penicillin was a remarkable achievement, the battle against infectious diseases was ongoing.

In addition to its immediate health benefits, penicillin also had far-reaching effects on society and economies. Healthier populations meant more productive communities. People who once succumbed to infections could now contribute to their families and societies. It's hard to quantify the economic impact, but countless stories reflect how families could flourish when they no longer lived under the constant threat of infectious diseases.

As we look at the current landscape of global health, penicillin remains a cornerstone in our medical toolkit. It laid the groundwork for understanding antibiotics and their use in treating various infections. Even today, the principles of antibiotic therapy are rooted in the lessons learned from penicillin's history.

Penicillin's impact on global health is a testament to the power of science, innovation, and human resilience. It has woven itself into the fabric of healthcare, and its legacy continues to inspire generations of medical professionals. In sharing these stories, we recognize that every breakthrough carries with it the potential to change lives, bridging the gap between suffering and healing. So the next time you hear about antibiotics, remember that behind the science lies a rich tapestry of human experience, triumphs, and ongoing challenges in the quest for health and well-being.

3.1 Penicillin in WWII and Public Health Campaigns

When we think of World War II, images of bravery, sacrifice, and battlefields often come to mind. But tucked away from the front lines, a quiet revolution was happening in medicine—one that would change the landscape of public health forever. Penicillin, the world's first true antibiotic, became a game-changer during this tumultuous period, significantly impacting soldiers and civilians alike.

Picture a soldier, weary and worn from battle, returning from the front lines only to fall ill to an infection. In the early days of the war, bacterial infections were as deadly as enemy fire. In fact, more soldiers died from infections than from combat wounds. The grim reality was that a small wound could turn fatal if untreated. It was against this backdrop that penicillin emerged, almost like a miracle.

The story begins with a race against time. As reports of infection-related deaths mounted, researchers ramped up efforts to mass-produce penicillin. Scientists like Howard Florey and Ernst Boris Chain worked tirelessly to refine penicillin extraction methods, making it more accessible for widespread use. Their efforts led to penicillin being manufactured in large quantities just in time for the D-Day invasion in 1944. Imagine the relief felt by medical staff as they treated wounded soldiers with this newfound miracle drug. What was once a life-threatening condition could now be treated effectively, allowing many to return home to their loved ones.

Let's take a moment to look at a real-life story. Dr. John McGrath, a combat medic during the war, often shared his experiences in interviews. He recalled treating soldiers who had suffered injuries during intense battles. One particular incident that stood out to him involved a young man who had been shot in the leg. Dr. McGrath administered penicillin, and within days, the soldier was on the mend. "It felt like I was watching a miracle unfold," he said. "We were able to save lives that would have otherwise been lost to infections."

But the impact of penicillin extended beyond the battlefield. As soldiers returned home, they brought with them not only stories of heroism but also a new understanding of health and hygiene. Public health campaigns emerged in the wake of the war, focusing on educating communities about the importance of sanitation and disease prevention. The successes of penicillin during WWII paved the way for governments and health organizations to promote antibiotics as essential tools in the fight against infections.

In the years that followed, public health campaigns began to emphasize the importance of early treatment and the responsible use of antibiotics. These campaigns aimed to educate the population about recognizing the signs of infections and seeking medical help promptly. The message was clear: just as soldiers needed penicillin on the battlefield, civilians also needed it in their daily lives.

As penicillin became more widely available, it was often viewed as a miracle drug that could cure anything. However, this perception also led to challenges. In some cases, individuals began to misuse antibiotics, expecting quick fixes without proper medical guidance. Public health officials quickly recognized this trend and sought to educate the public about the importance of using antibiotics responsibly. The narrative shifted from simply providing access to medications to promoting understanding about their appropriate use.

The legacy of penicillin in WWII is not just about the drug itself but also about the lessons learned in public health and medical treatment. It laid the foundation for the antibiotic revolution, inspiring a wave of research that led to the development of numerous other antibiotics. This era demonstrated the profound connection between medicine and public health campaigns, emphasizing that when armed with knowledge, communities could combat disease more effectively.

As we reflect on this significant chapter in history, it's essential to recognize that penicillin's impact went beyond its immediate medical benefits. It transformed public perceptions of health, emphasized the need for hygiene, and highlighted the importance of responsible antibiotic use. The stories of soldiers, doctors, and public health officials all intertwine to create a narrative of hope and resilience that still resonates today.

In a world where we sometimes take antibiotics for granted, remembering their history helps us appreciate the battles fought—both on the front lines and in the realm of public health. The legacy of penicillin continues to remind us of the incredible power of science, collaboration, and human spirit in overcoming challenges, both during wartime and beyond.

3.2 Reduction in Mortality Rates for Infectious Diseases

When penicillin burst onto the medical scene in the 1940s, it felt like a new dawn had broken for humanity. The story of its impact is not just about a groundbreaking drug; it's a narrative of hope, healing, and the dramatic transformation of public health.

Before penicillin, a simple infection could spiral into a life-threatening condition. Imagine a world where a minor cut or scrape could lead to severe complications. Stories from that era reveal the grim realities many faced. A young child in a small town might scrape their knee while playing outside. Instead of returning home with a few band-aids, the family would worry about the possibility of tetanus or a raging infection. This fear was all too real and often justified, leading to countless preventable deaths.

The introduction of penicillin marked a turning point. In the years following its mass production, medical professionals began to see a dramatic shift in patient outcomes. Hospitals that once struggled to manage infections now experienced fewer complications. It was not just a matter of statistics; it was about lives saved and families reunited.

Take the story of a man named Frank, who lived through the early days of penicillin. In his youth, Frank suffered from pneumonia, a disease that claimed many lives in those days. "I remember my mother's fear as I lay in bed, battling a fever that made me feel like I was in a waking nightmare," he shared in an interview decades later. Fortunately, by the time Frank fell ill, penicillin had become available. He was treated with this revolutionary drug and made a full recovery. Frank often reflected on how penicillin saved not just his life but also countless others, allowing them to experience moments they might have missed—birthdays, graduations, and family gatherings.

The reduction in mortality rates for infectious diseases was astounding. Diseases that once swept through communities with little warning were suddenly manageable. According to historical data, deaths from infections like pneumonia and syphilis plummeted in the decades following the introduction of antibiotics. By the 1950s, the impact of penicillin was so profound that health experts noted a significant decline in overall mortality rates due to infectious diseases.

But the story doesn't end there. As penicillin paved the way, it opened the door to the development of other antibiotics, creating a powerful arsenal in the fight against bacteria. Each new antibiotic added another layer of protection against diseases that had plagued humanity for centuries. The focus shifted from merely treating infections to preventing them entirely, leading to improvements in overall public health.

Public health campaigns embraced this newfound hope, educating communities on hygiene practices and the importance of seeking medical attention early. People began to understand that infections could be treated effectively, reducing the stigma associated with illness and encouraging individuals to seek help without fear.

However, this incredible journey also comes with a cautionary tale. With the widespread use of antibiotics, we began to encounter new challenges, such as antibiotic resistance. As we celebrated the reduction in mortality rates, we also learned that we needed to use these powerful tools responsibly. This lesson has become central to public health discussions today, reminding us that our victories are often accompanied by new battles.

Reflecting on the journey of penicillin, we can see how far we've come. The stories of lives saved, families reunited, and communities thriving paint a vivid picture of the transformative power of medicine. Frank's story is just one of many that highlight the hope that penicillin brought to countless individuals.

In our contemporary world, where antibiotics are common, it's easy to forget the struggles of the past. But understanding this history is crucial as we navigate modern health challenges. By honoring the legacy of penicillin and recognizing its role in reducing mortality rates, we can appreciate the significance of responsible antibiotic use and the ongoing need for innovation in medicine.

As we look ahead, let's carry forward the lessons learned from penicillin's journey. The fight against infectious diseases is ongoing, and while we have made remarkable strides, we must remain vigilant in safeguarding the health of future generations.

Penicillin was more than a medical breakthrough; it was a beacon of hope that illuminated a path toward a healthier world.

3.3 Penicillin's Role in Modern Healthcare Systems

In the world of modern medicine, penicillin remains a cornerstone, a reliable ally in our ongoing battle against infectious diseases. Its journey from discovery to daily practice is a testament to its enduring significance and the profound impact it has had on healthcare systems around the globe.

Imagine stepping into a bustling hospital today. Doctors and nurses dart through the hallways, their minds racing to address a myriad of patient needs. In one corner, a child with an ear infection is being treated with penicillin. In another, an elderly patient is receiving pneumonia. Though it may seem routine now, these scenarios are grounded in a remarkable history that has shaped the very fabric of our healthcare systems.

Penicillin's role in modern healthcare goes far beyond its antibiotic properties; it symbolizes hope and resilience. Consider the story of Sarah, a nurse who has dedicated her life to helping others. Early in her career, she encountered a young boy suffering from a severe strep infection. The boy was gravely ill, and his family was frightened. Sarah remembered the training she had received about penicillin's effectiveness in treating such infections. She administered the antibiotic, and within days, the boy's condition improved dramatically. "I'll never forget the look on his mother's face when she saw him smile again," Sarah recounted. "It reminded me why I became a nurse in the first place."

Beyond individual stories, penicillin has a wider impact on public health. Its introduction allowed for more effective treatment protocols and laid the groundwork for the development of more advanced antibiotics. This evolution has led to a significant decrease in hospitalization times and improved patient outcomes. Healthcare systems can allocate resources more efficiently, knowing they have a reliable weapon against bacterial infections.

Penicillin also plays a crucial role in surgical procedures. Surgeons rely on antibiotics to prevent infections during and after operations. The fact that a patient can undergo a complex procedure today, with a lower risk of postoperative infection thanks to antibiotics like penicillin, speaks volumes about the advancements in medical practices. The presence of penicillin has transformed what used to be risky endeavors into more routine procedures.

However, the journey has not been without challenges. As healthcare systems integrate penicillin into treatment protocols, they must also contend with the rise of antibiotic resistance. It's a complex issue that has arisen partly due to the overuse of antibiotics. Healthcare providers are now more vigilant, advocating for responsible use of antibiotics while continuing to rely on penicillin's efficacy when appropriate. The challenge lies in educating both medical professionals and patients about the importance of using antibiotics judiciously.

Moreover, the global health landscape has shifted, revealing disparities in access to medications, including penicillin. In many parts of the world, antibiotics remain difficult to obtain. This inequity emphasizes the need for ongoing efforts to ensure that penicillin and other essential medications reach those who need them most. Organizations are working tirelessly to improve access, recognizing that no one should be left vulnerable to treatable infections.

As we consider the role of penicillin in modern healthcare, it's crucial to acknowledge its place within the broader context of medical history. Penicillin was not merely a breakthrough; it inspired generations of scientists and healthcare professionals to seek out new treatments and innovations. It paved the way for research that continues to yield new antibiotics and therapies, ensuring that our healthcare systems remain robust in the face of evolving challenges.

Looking ahead, we can draw valuable lessons from penicillin's journey. The principles of discovery, responsible use, and equitable access will guide the future of healthcare. By celebrating penicillin's contributions and addressing the challenges it presents, we can continue to honor its legacy while striving for a healthier world for all.

In every corner of modern healthcare systems, from bustling hospitals to community clinics, penicillin serves as a reminder of what is possible when science, compassion, and resilience come together. It embodies the hope that with each dose, we can conquer infections and improve the lives of countless individuals, just as it has done for decades.

Chapter 4

Penicillin Resistance

As we navigate the landscape of modern medicine, one pressing issue has emerged like a shadow over the progress we've made in treating infections: penicillin resistance. This phenomenon is not just a scientific concern; it's a story of evolution, adaptation, and the ongoing battle between bacteria and antibiotics that plays out in real life every day.

To understand the gravity of penicillin resistance, it helps to consider the journey of Tom, a middle-aged man who once thought he had a simple infection. He had a sore throat and fever, so he visited his doctor, who prescribed penicillin. Tom felt relieved; he had heard about the miraculous effects of penicillin and how it could cure many infections. But after a few days, he noticed that his symptoms weren't improving. Confused and concerned, he returned to the doctor, who ran some tests and delivered the troubling news: the bacteria causing Tom's infection had developed resistance to penicillin.

Tom's experience illustrates a growing reality in healthcare. While penicillin has been a trusted ally against bacterial infections since its discovery, the bacteria themselves are continually evolving. They adapt, learning how to survive the very antibiotics designed to eliminate them. This resistance can stem from several factors, including overuse and misuse of antibiotics in both healthcare and agriculture, which gives bacteria more opportunities to evolve. Each time someone takes antibiotics unnecessarily, it's like giving bacteria a lesson on how to become stronger and more resilient.

Anecdotes abound in the medical community about patients facing penicillin-resistant infections. Dr. Emily, an infectious disease specialist, recalls a particularly challenging case involving a young woman named

Lisa. After undergoing surgery, Lisa developed a severe infection that should have been easily treatable with penicillin. However, the lab results showed that the bacteria were resistant. "It was frustrating," Dr. Emily said. "We had to resort to a much stronger antibiotic with more potential side effects, and it felt like we were taking a step back in our treatment approach."

Such experiences are becoming increasingly common, leading to a profound sense of urgency among healthcare professionals. They recognize that while antibiotics like penicillin have saved countless lives, their effectiveness is not guaranteed. The rise of resistance has prompted a shift in how doctors prescribe antibiotics, emphasizing the importance of only using them when absolutely necessary. This not only helps combat resistance but also preserves the effectiveness of these crucial medications for future generations.

The impact of penicillin resistance extends beyond individual patients. It poses a significant challenge to public health systems worldwide. Hospitals are grappling with higher rates of infections that are harder to treat, leading to longer hospital stays and increased healthcare costs. The World Health Organization has labeled antibiotic resistance one of the biggest threats to global health, urging countries to take decisive action.

But it's not all doom and gloom. There are signs of hope as well. Scientists are working tirelessly to develop new antibiotics and alternative treatments, looking for innovative ways to combat resistant bacteria. Some researchers are exploring phage therapy, which uses viruses that specifically target bacteria. Others are looking into vaccines that could prevent bacterial infections altogether, reducing the reliance on antibiotics in the first place.

Education and awareness are also critical in this fight. Campaigns to promote responsible antibiotic use are gaining traction, encouraging patients and healthcare providers alike to be mindful of when and how antibiotics are prescribed. When people understand the consequences of overusing antibiotics, they are more likely to adhere to prescribed treatments and advocate for appropriate medical care.

Ultimately, the story of penicillin resistance is a reminder of the delicate balance between medical advances and the biological realities of life. It's a dynamic relationship that requires continuous learning and adaptation. By fostering collaboration between researchers, healthcare providers, and patients, we can work together to mitigate the impacts of resistance and ensure that antibiotics like penicillin continue to serve their life-saving purpose.

As we look to the future, it's essential to embrace both the challenges and the solutions. The journey of penicillin and its role in our lives is far from over. By respecting the power of antibiotics and treating them as precious tools rather than casual remedies, we can honor the legacy of penicillin and protect the health of our communities for years to come.

4.1 Understanding Antibiotic Resistance

Antibiotic resistance is a term that seems to be on everyone's lips these days, yet it often remains shrouded in mystery for many. To truly grasp its significance, we need to peel back the layers and stays, increased medical costs, and higher mortality rates. The World Health Organization has labeled antibiotic resistance a global health crisis, urging governments to take action to combat this escalating threat.

But while the statistics and science behind antibiotic resistance can seem daunting, there is hope. Awareness is growing, and efforts are being made to tackle the problem on multiple fronts. Education campaigns are helping both healthcare providers and the public understand the importance of responsible antibiotic use. By ensuring antibiotics are prescribed only when necessary and following the prescribed regimen, we can reduce the chance of resistance developing.

In the lab, researchers are tirelessly working to develop new antibiotics and alternative treatments to stay one step ahead of resistant bacteria. Some are investigating natural compounds, while others are exploring innovative methods like bacteriophage therapy, which uses viruses to target and kill bacteria.

Public health initiatives are also focusing on prevention, promoting vaccinations and hygiene practices to reduce the need for antibiotics in the first place. For instance, the flu vaccine not only protects individuals from getting sick but also helps minimize the number of people seeking antibiotics for secondary infections.

In understanding antibiotic resistance, we are reminded that it is not just a scientific issue but a human one. It affects our friends, families, and communities. By sharing stories, raising awareness, and advocating for responsible antibiotic use, we can contribute to a collective effort to combat this challenge. It's about recognizing that antibiotics are valuable tools that must be used wisely.

So, as we navigate our health journeys—whether it's dealing with a common cold or a more serious infection—let's remember the lessons of the past. By honoring the power of antibiotics and treating them with the respect they deserve, we can help ensure that future generations can benefit from the life-saving properties of these incredible medicines.

4.2 How Bacteria Develop Resistance to Penicillin

Understanding how bacteria develop resistance to penicillin can feel like peering into a secret world where tiny organisms outsmart powerful medications. To grasp this concept, let's dive into the fascinating journey of bacteria and their uncanny ability to adapt.

Imagine a bustling city, alive with people, cars, and buildings. Now, envision that city under threat from a powerful storm. As the storm approaches, some people might seek shelter in their homes, while others might find alternative routes to safety. Similarly, bacteria face their own challenges when confronted with antibiotics like penicillin, and they have developed remarkable strategies to survive.

When penicillin was first introduced, it was a game-changer in the fight against infections. However, bacteria are clever. They can undergo mutations or share genetic material with one another, allowing them to create defenses against antibiotics. This ability to adapt is a survival mechanism, honed through millions of years of evolution.

One fascinating story comes from a researcher named Dr. Kim, who spent years studying a particular strain of bacteria known for its resilience. "I remember the first time I isolated a resistant strain in the lab," she recalls. "It was a real eye-opener. I had been treating these bacteria with penicillin for weeks, and yet they just kept growing."

So, how exactly do these bacteria develop resistance? Let's break it down. One common method is through genetic mutations. Just like humans, bacteria can have random changes in their DNA. Sometimes, these mutations can be beneficial. For instance, if a bacteria mutates in such a way that it can produce an enzyme called beta-lactamase, it can break down penicillin before it has a chance to do any damage. This newfound ability allows the bacteria to thrive despite the presence of the antibiotic.

Another way bacteria can become resistant is through horizontal gene transfer. This is a process where bacteria can exchange genetic material with their neighbors. Think of it like a high school where students trade notes before a big exam. One bacteria might have a gene that allows it to resist penicillin, and when it meets another bacteria, it can share that information. Before you know it, a whole community of bacteria can be resistant, making infections much harder to treat.

This phenomenon has been particularly troubling in healthcare settings. Hospitals can be breeding grounds for resistant bacteria, especially when antibiotics are overprescribed. A nurse named Maria shared her experience working in a busy ICU. "We'd often see patients with infections that didn't respond to the usual treatments," she says. "It was frustrating and sometimes heartbreaking because we had to resort to stronger, more toxic drugs that came with serious side effects."

The impact of antibiotic resistance is not just a clinical concern; it can affect the entire healthcare system. When standard treatments fail, patients may need longer hospital stays or even face life-threatening situations. It's a sobering reality, highlighting the importance of understanding how resistance develops.

However, not all is lost. Awareness is increasing, and there's a growing movement towards responsible antibiotic use. Educating both healthcare professionals and patients about the dangers of misuse is crucial. When antibiotics are prescribed only when necessary, and patients complete their full courses of treatment, we can help slow down the spread of resistance.

Moreover, scientists are exploring new avenues for combating resistant bacteria. Research is being conducted on alternative treatments, such as bacteriophage therapy, where viruses that specifically target bacteria are utilized. This innovative approach could offer a new weapon in our fight against resistant infections.

In understanding how bacteria develop resistance to penicillin, we recognize the resilience of life in all its forms. It's a dance of survival, adaptation, and discovery. By sharing stories and knowledge, we can empower ourselves and our communities to be part of the solution, ensuring that antibiotics remain effective tools in our medical arsenal. After all, it's not just about defeating infections; it's about preserving the power of antibiotics for future generations.

4.3 The Rise of Resistant Strains: MRSA and More

When we think of the advancements in medicine, it's hard not to feel a sense of awe. Just a century ago, simple infections could be life-threatening. Enter penicillin, and the world of antibiotics opened up new possibilities for treating diseases. However, alongside this progress, we've witnessed the rise of a formidable opponent: antibiotic-resistant strains of bacteria, with MRSA (Methicillin-Resistant Staphylococcus Aureus) being one of the most notorious.

Let's rewind to a time not so long ago when MRSA emerged in hospitals. A nurse named Lisa vividly recalls her early experiences with this superbug. "I remember my first encounter with a patient who had MRSA. It was a tough situation. We were all on high alert because we knew how challenging these infections could be," she shared. MRSA is a strain of Staphylococcus aureus that has evolved to resist methicillin, a common antibiotic. While the bacteria itself can be found on our skin and in our noses without causing harm, when it invades deeper into the body, it can lead to severe infections, especially in vulnerable patients.

The story of MRSA is one of adaptation and survival. When antibiotics like methicillin became widely used in the 1960s, bacteria began to develop resistance. It's almost like a game of chess, where bacteria continuously strategize against the medications designed to eliminate them. Instead of surrendering, they find ways to outsmart the drugs, leaving healthcare professionals grappling with limited treatment options.

In recent years, MRSA has also made its way into the community, affecting otherwise healthy individuals. This shift surprised many experts. "At first, we thought MRSA was strictly a hospital problem," Dr. Sarah, an infectious disease specialist, explains. "But then we started seeing cases in schools, gyms, and even homes." The reality hit hard when community-acquired MRSA began infecting athletes and kids playing sports, highlighting how easily bacteria can spread in everyday environments.

In response to the rising tide of resistant strains, scientists and public health officials are fighting back with renewed vigor. The focus has shifted toward educating the public about proper hygiene practices. Simple measures like regular handwashing, keeping wounds covered, and not sharing personal items can go a long way in preventing the spread of MRSA and other resistant strains.

Yet, the story doesn't end with MRSA. Other resistant bacteria, like VRE (Vancomycin-Resistant Enterococcus) and CRE (Carbapenem-Resistant Enterobacteriaceae), are also making headlines. Each of these resistant strains carries its own set of challenges. Take VRE, for example. Often found in the intestines, it can cause serious infections in patients with weakened immune systems. Similarly, CRE is known as a "nightmare bacteria" due to its resistance to nearly all available antibiotics.

These developments have made the fight against infections more complex. "It's like we're in a constant race against time," says Dr. Samuel, an epidemiologist. "For every new antibiotic developed, it feels like bacteria find a way to counteract it." The urgency to discover new treatments and strategies has never been more pronounced.

Despite the challenges, there are stories of hope. Research is being conducted into alternative therapies, such as phage therapy, which uses viruses that specifically attack bacteria. This approach is still in its infancy, but it represents a promising avenue in the ongoing battle against antibiotic resistance.

As we navigate this landscape, it's important to remember the role we each play. Responsible antibiotic use is crucial. Patients should complete their prescribed courses and never use antibiotics for viral infections like the common cold. Additionally, healthcare providers are encouraged to prescribe antibiotics only when truly necessary.

The rise of resistant strains like MRSA serves as a reminder of the delicate balance in our healthcare system. It's a complex interplay between medical advancements and the resilience of bacteria. By sharing stories and fostering understanding, we can empower ourselves and our communities to take meaningful steps toward combating antibiotic resistance. After all, this battle is not just about science; it's about protecting our health and the health of future generations.

4.4 Strategies to Combat Antibiotic Resistance

Antibiotic resistance is a complex and pressing issue that requires collective effort and innovative strategies to tackle. As we find ourselves in a world where once-treatable infections are becoming increasingly resistant to standard antibiotics, it's vital to explore ways we can all contribute to combating this growing threat. Let's dive into some effective strategies that can make a real difference.

One of the most crucial steps in addressing antibiotic resistance begins with understanding how we use antibiotics. Take Sarah, a concerned mother, for instance. A few years ago, her daughter came down with a fever and a cough. Panicked, Sarah rushed to the pediatrician, convinced her child needed antibiotics. However, the doctor calmly explained that it was likely a viral infection and that antibiotics wouldn't help. This experience taught Sarah a valuable lesson: not all illnesses require antibiotics. Responsible prescribing is key. Healthcare providers are now more aware than ever of the importance of only prescribing antibiotics when absolutely necessary. By ensuring that these powerful medications are reserved for bacterial infections, we can slow down the development of resistance.

Education plays a pivotal role as well. It's not just about doctors; patients also need to understand the implications of their treatment choices. Community outreach programs have emerged, aiming to inform people about the dangers of antibiotic misuse. Local health departments have organized workshops where families can learn about the importance of hygiene and proper antibiotic usage. For instance, during a community fair, a pharmacist named James set up a booth to educate attendees about the difference between viral and bacterial infections. "I was surprised by how many people didn't realize that antibiotics wouldn't work on colds or the flu," he shared. By breaking down these misconceptions, we empower individuals to make informed decisions about their health.

In the healthcare setting, infection prevention measures are crucial. Hospitals are on the front lines in the battle against resistant bacteria. Implementing rigorous hand hygiene protocols and using personal protective equipment can significantly reduce the spread of infections. Nurses, like Anna, play a vital role in this effort. "I make it a point to remind my colleagues about the importance of handwashing between patients. It might seem basic, but it's one of the most effective ways to prevent infections," she explains. By fostering a culture of safety and accountability, healthcare professionals can create an environment where antibiotic resistance has less opportunity to thrive.

Another promising avenue is the exploration of alternative treatments. Scientists are delving into phage therapy, a method that uses bacteriophages—viruses that specifically infect bacteria—to target resistant strains. While still largely in the experimental stage, phage therapy has shown potential in treating infections that have resisted traditional antibiotics. Dr. Laura, a microbiologist working on this frontier, shares her excitement: "It's fascinating to think about harnessing natural predators of bacteria to combat infections. We're just scratching the surface, but the possibilities are endless." This approach underscores the importance of ongoing research and innovation in our fight against antibiotic resistance.

Vaccination is another powerful strategy. By preventing infections in the first place, we can reduce the need for antibiotics. Think about the impact of the pneumonia vaccine. It has significantly decreased the incidence of pneumococcal infections, which previously required extensive antibiotic treatment. Communities that embrace vaccination campaigns, like the one organized by local health advocates, are taking proactive steps to protect their members from infections that could lead to antibiotic use.

Lastly, we can't underestimate the importance of global collaboration. Antibiotic resistance knows no borders, and solutions must be sought on an international scale. Organizations like the World Health Organization (WHO) are working to develop global action plans, promoting awareness and research funding to combat resistance worldwide. A vivid example comes from a recent conference where experts from around the globe gathered to share strategies and successes. The enthusiasm was palpable as they discussed how pooling resources and knowledge could lead to innovative solutions.

Combating antibiotic resistance requires a multifaceted approach that involves education, responsible use, infection prevention, alternative treatments, vaccination, and global collaboration. Each of us has a role to play, whether it's making informed decisions about our health, advocating for responsible prescribing, or participating in community education initiatives. The journey may be challenging, but as we work together, we can protect the effectiveness of antibiotics for future generations. Remember, the fight against antibiotic resistance is not just a battle for scientists and doctors; it's a shared mission that starts with you and me.

Chapter 5

Side Effects and Risks

When we think about antibiotics, we often focus on their life-saving properties. Penicillin, for instance, has been a cornerstone in treating bacterial infections for decades. However, like any medication, it comes with its own set of side effects and risks that are important to understand. Navigating these potential pitfalls is key to ensuring that we use antibiotics effectively and safely.

Consider the story of Mike, a young man who had been feeling under the weather for weeks. After a visit to his doctor, he was prescribed penicillin to treat a persistent throat infection. At first, Mike was relieved;

he thought he was on the road to recovery. But a few days into his treatment, he noticed a strange rash spreading across his arms. Confused and worried, he called his doctor, who informed him that this could be a common allergic reaction to penicillin. For Mike, this was an unexpected twist in his journey to health. Thankfully, his doctor was able to switch him to a different antibiotic, but it was a stark reminder that even widely-used medications can have unexpected effects.

Allergic reactions are perhaps the most talked-about side effect of penicillin. They can range from mild symptoms, like rashes or itching, to more severe reactions, such as anaphylaxis—a life-threatening condition that requires immediate medical attention. Many people might not realize that they can develop an allergy to penicillin even if they've taken it before without any issues. That's why it's essential for healthcare providers to take a thorough medical history and for patients to communicate any previous experiences with antibiotics.

Another common side effect associated with penicillin and many other antibiotics is gastrointestinal upset.

Some patients, like Jane, a busy mother of three, found this out the hard way. After starting her penicillin course, she experienced nausea and diarrhea that made her feel even worse than her original infection. "It was frustrating because I was trying to get better, but the antibiotics left me feeling miserable," she recalls. While these side effects can be unpleasant, they are usually temporary and resolve once the course of antibiotics is finished.

Antibiotics can also disrupt the natural balance of bacteria in our bodies, particularly in the gut. This can lead to issues like Clostridium difficile (C. diff) infections, which can cause severe diarrhea and more serious complications. Doctors have increasingly recognized this risk, especially in older adults or those with weakened immune systems. It's a reminder that while antibiotics are powerful tools, they should be used judiciously.

In addition to allergic reactions and gastrointestinal issues, there are other risks to consider. For instance, using antibiotics like penicillin can sometimes lead to secondary infections. This occurs when the antibiotic wipes out not only the harmful bacteria causing the initial infection but also beneficial bacteria that help keep our bodies balanced. As a result, a patient might find themselves dealing with a new issue after completing their antibiotic regimen.

Understanding these side effects and risks highlights the importance of communication between patients and healthcare providers. It's crucial for patients to feel empowered to discuss any symptoms they experience during treatment. When Mike called his doctor about the rash, it not only helped him but also allowed the doctor to note this allergy for future reference. This open dialogue can lead to more personalized and safer treatment plans.

Moreover, the potential for antibiotic resistance also looms over the conversation about side effects. Each time we use antibiotics, we risk contributing to the larger problem of resistant bacteria. This is especially true if the antibiotics are misused or overused, which can result in longer treatment times and increased side effects.

So, how can we navigate the landscape of side effects and risks associated with penicillin? Education is key. Patients should ask questions and understand the medication they're taking. It's helpful to discuss what to expect, both in terms of benefits and potential side effects.

While penicillin has transformed the treatment of bacterial infections, being aware of its side effects and risks is crucial for safe and effective use. It's a balancing act—understanding the potential pitfalls while also appreciating the tremendous benefits antibiotics offer. Just like Mike and Jane, many patients can share their stories, underscoring the need for a cautious yet informed approach. By fostering a culture of communication and education around antibiotic use, we can ensure that penicillin continues to save lives without unnecessary complications.

5.1 Common Side Effects of Penicillin

When we think of penicillin, it's often with gratitude for its ability to save lives. Since Alexander Fleming discovered it in 1928, penicillin has become a beacon of hope for treating bacterial infections. However, just like any medication, it can come with its share of side effects. Understanding these side effects can make the difference between a successful treatment and a frustrating experience.

Take Sarah, for instance. She was feeling under the weather for days, battling a stubborn ear infection that left her fatigued and achy. When her doctor prescribed penicillin, she felt a wave of relief wash over her. After a few days, however, she noticed something unsettling: a rash had started to appear on her arms. "I didn't think much of it at first, but as the days went on, it spread," she recalled. It turned out that Sarah had developed an allergic reaction to penicillin, something that can happen even to those who have taken it before without any issues.

Allergic reactions are among the most common side effects associated with penicillin. They can manifest in various ways, from mild symptoms like rashes and itching to more severe reactions such as difficulty breathing and anaphylaxis. Anaphylaxis is rare but can be life-threatening, requiring immediate medical attention. This underscores the importance of informing healthcare providers about any past reactions to antibiotics or known allergies.

Aside from allergic reactions, gastrointestinal issues often accompany penicillin treatment. Many patients, including Sarah, experience symptoms like nausea, diarrhea, and stomach cramps. It's not uncommon for people to feel a bit queasy during their course of antibiotics. This was true for David, who had been prescribed penicillin for a dental infection. He found himself racing to the bathroom more often than he'd like, which added to his discomfort. "I thought I was getting better, but then the side effects hit me," he said, laughing about it now. While these side effects can be unpleasant, they often resolve once the antibiotic course is completed.

Another important aspect to consider is how penicillin can disrupt the balance of bacteria in our bodies. Our gut is home to a complex ecosystem of beneficial bacteria that help with digestion and overall health. Antibiotics, including penicillin, can disturb this balance, sometimes leading to secondary infections like Clostridium difficile (C. diff). This can result in severe diarrhea and additional complications. This is why healthcare providers are increasingly cautious when prescribing antibiotics, especially to older adults or those with compromised immune systems.

The risk of yeast infections is another common side effect that can occur when taking penicillin. Women, in particular, may experience this as the antibiotic affects the natural flora of the body. Jenna, a college student, learned this first hand after completing her course of penicillin for a sinus infection. She felt relieved to be rid of the infection but soon found herself dealing with an unwelcome yeast infection. "It was the last thing I expected," she chuckled, sharing her experience. "It was like trading one problem for another."

It's essential for patients to be aware of these common side effects and to have open conversations with their healthcare providers. Understanding what to expect can make a world of difference. When Sarah noticed her rash, she promptly called her doctor, who advised her to stop taking the medication and switch to a different antibiotic. This swift action helped her avoid further complications.

While penicillin is a powerful tool in fighting infections, it's important to acknowledge its potential side effects. By sharing stories like those of Sarah, David, and Jenna, we can humanize the experience of taking antibiotics. Awareness of these common side effects not only prepares patients but also fosters a collaborative approach to healthcare. When patients feel informed and empowered, they are more likely to communicate openly with their healthcare providers, ensuring a safer and more effective treatment journey.

5.2 Allergic Reactions: Symptoms and Management

When it comes to antibiotics like penicillin, most of us associate them with healing. However, the flip side of this miracle drug can sometimes be a set of unexpected challenges. Allergic reactions to penicillin, while not overly common, can be serious and sometimes frightening. Understanding these reactions, their symptoms, and how to manage them can make a significant difference in your treatment journey.

Imagine Anna, a young woman who was always careful about her health. She had heard countless stories about antibiotics saving lives and was relieved when her doctor prescribed penicillin for a sinus infection. After a few doses, she felt better—but then, she noticed an itch creeping up her arms. At first, she shrugged it off as a minor annoyance. But as the days went by, that little itch turned into a full-blown rash. "I thought it was just a reaction to the change in the weather," she recalls. Unfortunately, it wasn't. Anna was experiencing an allergic reaction to penicillin.

So, what exactly does an allergic reaction to penicillin look like? The symptoms can vary widely, and they often begin within hours to days after taking the medication. Common signs include rashes, hives, itching, and swelling, particularly around the face and throat. Some people, like Anna, may also experience gastrointestinal issues like nausea or diarrhea. However, in more severe cases, individuals can develop anaphylaxis—a life-threatening reaction characterized by difficulty breathing, rapid swelling, and a drop in blood pressure.

Recognizing the symptoms is crucial. If you notice any of these signs after starting penicillin, it's essential to act quickly. For Anna, the rash worsened, and she found herself feeling dizzy. Thankfully, she remembered her doctor's advice: if anything felt off, don't hesitate to seek help. She called her doctor's office and was told to stop the medication immediately and come in for an evaluation. This quick thinking likely prevented a more severe situation.

For anyone experiencing these symptoms, it's important to know that you're not alone. Many people have gone through similar experiences, and there's a wealth of support available. Healthcare providers are trained to handle allergic reactions, and they often recommend alternatives if penicillin is deemed unsuitable for you. In Anna's case, her doctor prescribed a different antibiotic that worked just as effectively without the unwanted side effects.

Managing allergic reactions involves more than just switching medications. It's about prevention, awareness, and having a plan. If you've had a previous allergic reaction to penicillin, inform your healthcare provider before starting any new treatment. Keeping a record of your allergies can be incredibly helpful. Many people carry a card in their wallets or wear medical alert bracelets to ensure that anyone providing care knows about their allergies.

Education is another key piece of the puzzle. The more you know about your body's reactions, the better equipped you'll be to handle any surprises. For instance, understanding that allergic reactions can sometimes get worse with each exposure can encourage vigilance. Anna, now more aware of her body's responses, regularly discusses her allergies with her healthcare providers, ensuring they have a complete picture of her medical history.

It's also important to have a support system in place. Friends and family should know what to watch for if you're taking penicillin or any other antibiotic. If you ever find yourself in a situation where you're feeling unwell, don't hesitate to reach out for help. Like Anna, it's always better to err on the side of caution.

Allergic reactions to penicillin can be distressing, but they can also be managed effectively with knowledge and communication. By sharing stories of people like Anna, we can demystify the process and help others feel more prepared and supported. Awareness of symptoms, prompt action, and open dialogue with healthcare providers can turn what might be a scary experience into an opportunity for learning and growth. Remember, your health is a partnership between you and your medical team, and together, you can navigate any challenges that arise.

5.3 Who Should Avoid Penicillin?

When it comes to penicillin, it's easy to think of it as a universal cure-all. After all, this antibiotic has saved countless lives since its discovery. However, it's important to recognize that penicillin isn't suitable for everyone. Understanding who should avoid it can help prevent complications and ensure that individuals receive the most effective treatment for their specific needs.

Let's start with Sarah, a lively teenager who was thrilled to finally get rid of that stubborn ear infection. Her doctor prescribed penicillin, which she had heard great things about. However, Sarah had a family history of allergies, and she had experienced hives from certain medications before. When she took her first dose of penicillin, she felt a rush of excitement at the prospect of relief—until she started itching and swelling. Luckily, her mom recognized the signs of an allergic reaction and rushed her back to the doctor.

Sarah's experience highlights one of the most critical groups that should avoid penicillin: individuals with a known allergy to the drug or its derivatives. Allergic reactions can range from mild to severe, and for those who have experienced an anaphylactic reaction in the past, exposure to penicillin can be life-threatening. If you know you're allergic, it's vital to communicate this to your healthcare provider and ensure it's noted in your medical records.

But it's not just allergies we should consider. Some people may have certain medical conditions that make penicillin a risky choice. For instance, those with a history of severe asthma may be at higher risk for complications when taking penicillin, particularly if they also have a sensitivity to other medications. In such cases, alternative antibiotics may be prescribed to minimize the risk of respiratory issues.

Age can also play a significant role in determining whether someone should avoid penicillin. While many children can safely take it, infants under a certain age or those with specific health concerns may not be the best candidates. Likewise, the elderly population sometimes requires careful consideration. As our bodies change with age, the way we metabolize medications can shift, meaning that penicillin may not be as effective or could lead to more side effects.

Let's not forget about pregnant and breastfeeding women. While penicillin is generally considered safe during pregnancy, every individual is different. Factors like existing allergies or complications can change the risk-benefit ratio. For example, if a woman has had adverse reactions to penicillin in the past, doctors will likely look for alternatives to ensure both the mother and the baby are safe.

Consider Mike, a middle-aged man with a history of kidney issues. When he was prescribed penicillin for a bacterial infection, his doctor explained that his kidney function could be affected. The doctor decided to monitor him closely and adjust the dosage as necessary. This careful approach is crucial for individuals with renal impairment, as penicillin is processed through the kidneys, and reduced function can lead to an accumulation of the drug in the body.

For those with other underlying health issues, such as liver disease or certain autoimmune conditions, the same caution applies. Consulting with a healthcare provider who understands your complete medical history is essential to determine the best course of action. It's not just about treating the infection; it's about doing so in a way that considers the entire picture of your health.

While penicillin is a powerful tool in fighting infections, it's not always the right choice for everyone. By sharing stories like Sarah's, we can foster a better understanding of the factors that may necessitate avoiding this antibiotic. Always be open and honest with your healthcare provider about your medical history, allergies, and any other medications you're taking. Together, you can explore alternative treatments that suit your needs and ensure that you receive the best possible care. It's a collaborative journey toward health, and being informed is the first step in making that journey successful.

5.4 Interactions with Other Medications

When it comes to managing our health, many of us juggle multiple medications at once. Whether it's a daily vitamin, an occasional pain reliever, or a prescribed treatment for a chronic condition, the complex dance of medications can sometimes lead to unintended consequences. One antibiotic that plays a significant role in this mix is penicillin. Understanding how penicillin interacts with other medications is crucial for ensuring effective treatment and minimizing risks.

Let's think about Emily, a young woman who recently underwent surgery for a torn ligament. After her procedure, her doctor prescribed penicillin to prevent any potential infections. However, Emily was also taking blood thinners to manage her risk of clots. When she returned for a follow-up appointment, her doctor noticed that her blood levels were off. It turned out that penicillin had interfered with the effectiveness of her blood thinners, leading to increased levels of anticoagulation. This situation is a classic example of how interactions between medications can pose serious health risks.

One of the most common categories of drugs that penicillin interacts with is anticoagulants. These are medications designed to prevent blood clots, like warfarin. When penicillin is introduced into the system, it can enhance the effects of these blood thinners, increasing the risk of bleeding. This is why healthcare providers often monitor patients closely if they are prescribed both penicillin and an anticoagulant. In Emily's case, her doctor had to adjust her blood thinner dosage to keep her safe and effective throughout her recovery.

But interactions don't stop there. Penicillin can also play a role in how the body absorbs other medications. For instance, certain medications for heartburn or acid reflux, such as antacids, can bind to penicillin in the stomach, reducing its effectiveness. Picture Tom, an older gentleman who was prescribed penicillin for pneumonia. He often took his heartburn medication right after meals. Unfortunately, he didn't realize that doing so was lessening the power of the antibiotic. It wasn't until a follow-up visit that his doctor discovered his treatment was being compromised, leading to an extended illness.

Another important interaction occurs with certain oral contraceptives. While penicillin itself doesn't directly affect birth control pills, it can impact the gut flora responsible for metabolizing these medications. This can lead to decreased effectiveness of hormonal contraceptives, putting women at risk for unintended pregnancy. For this reason, women taking penicillin are often advised to consider using backup contraception during and after their antibiotic treatment.

Pain relievers, especially nonsteroidal anti-inflammatory drugs (NSAIDs) like ibuprofen, can also interact with penicillin. While these medications are commonly taken for discomfort, they can sometimes mask symptoms of a more serious infection. When Jack, a father of three, took ibuprofen alongside penicillin for a respiratory infection, he felt better almost immediately. However, this led him to overlook the persistent symptoms of his illness, delaying necessary follow-up care.

So how can you navigate these potential interactions? Communication is key. Before starting penicillin or any new medication, it's important to provide your healthcare provider with a complete list of everything you're taking, including over-the-counter drugs and supplements. Keeping a medication diary can be helpful, allowing you to track any changes and how they make you feel.

Additionally, it's wise to consult a pharmacist. They are a great resource for understanding medication interactions and can help you ensure that your treatments work harmoniously rather than in conflict. Don't hesitate to ask questions or seek clarification about how your medications interact.

Ultimately, understanding how penicillin interacts with other medications is about prioritizing your health. By being proactive and informed, you can help ensure that your treatment is as effective as possible. It's a partnership between you and your healthcare team, where open communication and attention to detail can lead to better outcomes. Just like in a well-choreographed dance, each step matters, and together you can create a rhythm that supports your journey toward recovery.

Chapter 6

Penicillin in Modern Medicine

When we think of modern medicine, it's easy to get lost in the flurry of high-tech treatments, robotic surgeries, and personalized medicine. Yet, beneath all that complexity lies a humble yet revolutionary discovery: penicillin. This antibiotic, discovered by Alexander Fleming in 1928, still stands as a cornerstone in our fight against infections. As we explore its role in modern medicine, we'll see not just its enduring power but also the stories of people whose lives were transformed because of it.

Take Sarah, for example, a vibrant young teacher who suddenly found herself battling a severe skin infection.

It started as a minor cut on her hand, but within days, it had swelled and turned red, leaving her feeling feverish and fatigued. After a quick visit to her doctor, Sarah was prescribed penicillin. Almost immediately, she noticed a difference. The swelling began to decrease, and her energy returned. Sarah's story is a testament to how penicillin continues to provide effective treatment for a variety of infections, from strep throat to pneumonia, even in today's fast-paced medical landscape.

Despite its long history, penicillin has evolved alongside our medical practices. Today, we have various forms and derivatives that are more effective against a broader range of bacteria. For instance, while the original penicillin G was primarily used to treat infections caused by gram-positive bacteria, newer variations can tackle resistant strains and different types of bacteria. This adaptability keeps penicillin relevant, allowing it to be a vital tool for doctors across the globe.

However, the effectiveness of penicillin isn't just about the drug itself; it's also about the systems surrounding its use. In many hospitals today, strict protocols exist to ensure antibiotics are prescribed wisely. This means not just reaching for penicillin at the first sign of trouble, but rather evaluating the situation thoroughly. Doctors often conduct cultures to identify the specific bacteria causing an infection. This careful approach minimizes the risk of developing antibiotic resistance, a growing concern in modern medicine.

In the past, the broad application of penicillin might have seemed like a miracle cure, leading to over-prescription and misuse. But today's healthcare providers are much more aware of the consequences. Penicillin's journey from a wonder drug to a measured, strategic treatment reflects a growing understanding of the delicate balance between efficacy and the potential for resistance.

Let's consider John, a father who was recently hospitalized for a serious infection. Initially, he was worried when the doctors mentioned penicillin; he had heard stories about allergies and complications. But through careful discussions with his healthcare team, John learned that while some people are allergic to penicillin, most can take it safely. His doctors monitored him closely, and soon, he found himself recovering quickly and returning to his family. This personalized approach to treatment is increasingly common in modern medicine, where the patient's experience and history are integral to medical decisions.

Moreover, penicillin is now part of a broader arsenal of antibiotics. While it may be the first antibiotic that comes to mind, it's often used in combination with other drugs for synergistic effects. For instance, during severe infections, a physician might prescribe a cocktail of antibiotics tailored to target different bacteria. This multi-faceted approach can enhance treatment efficacy and help combat resistant strains of bacteria that have emerged over the years.

In recent years, the global health community has recognized the importance of keeping penicillin at the forefront of antibiotic therapies. Programs aimed at increasing access to antibiotics in developing countries, where bacterial infections can be deadly due to a lack of treatment options, are essential. Through initiatives that promote the availability of penicillin and educate healthcare workers about its use, we're seeing a concerted effort to ensure this life-saving drug continues to make a difference in people's lives worldwide.

Penicillin's place in modern medicine is more than just a story of a drug; it's a narrative about resilience, innovation, and the enduring human spirit. It reminds us that sometimes, the most significant breakthroughs come from the simplest beginnings. In a world that can feel overwhelming with medical advancements, it's important to remember the story of penicillin: a story of hope, healing, and the unyielding pursuit of better health for all. Through the lens of patients like Sarah and John, we see that penicillin is not just a medication; it's a lifeline that continues to shape our approach to healthcare today and into the future.

6.1 New Applications and Combinations

As we continue to navigate the landscape of modern medicine, penicillin remains a relevant player, but its story doesn't end with the familiar battles against common infections. The world of antibiotics is continually evolving, and researchers are discovering new applications and combinations for penicillin that are transforming how we treat various conditions.

Let's explore some of these exciting developments, weaving in stories of how they're making a difference in real lives.

Take, for example, the innovative use of penicillin in treating certain chronic conditions. It might come as a surprise, but penicillin is being investigated for its potential role in managing conditions like cystic fibrosis. This genetic disorder, which affects the lungs and digestive system, leads to thick mucus buildup and frequent infections. Patients often find themselves in a vicious cycle of infections and treatments. Researchers are exploring how penicillin, in conjunction with other antibiotics, can help target the persistent bacteria that commonly invade the lungs of cystic fibrosis patients. This collaboration among antibiotics aims to enhance treatment effectiveness and improve quality of life.

Now, let's turn our attention to a remarkable story from a small town. There's Emma, a spirited teenager who had spent much of her life in and out of the hospital due to cystic fibrosis. Despite her challenges, she was always determined to keep up with her friends and pursue her dreams. Recently, her doctors decided to try a new combination therapy that included penicillin. They were hopeful that this approach would help reduce her lung infections.

With this new treatment, Emma began to notice a significant improvement. Her doctors were thrilled to see her lung function tests improve, and Emma felt more energetic than ever. The new application of penicillin, combined with other antibiotics, allowed her to engage in activities she had previously struggled to enjoy. Emma's story is a testament to the potential of penicillin beyond its traditional uses, showcasing how it can contribute to better health outcomes in chronic conditions.

Moreover, penicillin is being combined with other medications to enhance its effectiveness against resistant strains of bacteria. As antibiotic resistance has become a pressing global concern, researchers have turned their attention to finding ways to revive the power of older antibiotics like penicillin. One promising avenue is pairing penicillin with adjuvants—substances that help enhance its effects. For instance, some studies have indicated that combining penicillin with certain enzymes can improve its ability to penetrate bacterial defenses, making it effective against strains that have previously been resistant.

Imagine Mark, a middle-aged man who suffered from recurrent bacterial infections that left him feeling exhausted and frustrated. After numerous rounds of antibiotics that seemed ineffective, he felt like he was running out of options. His doctor introduced a new treatment protocol that included a combination of penicillin and an innovative adjuvant designed to boost the antibiotic's efficacy. Mark was amazed at how quickly he started feeling better. Within weeks, he was back at work and enjoying time with his family again. This experience illustrates how the innovative use of penicillin can lead to renewed hope for patients grappling with stubborn infections.

In addition to treating bacterial infections, researchers are also investigating how penicillin can be integrated into other fields of medicine. One fascinating area is its potential role in preventing infections in surgical settings. As we've learned from countless studies, the risk of postoperative infections can lead to significant complications. Some hospitals are now considering the use of penicillin as a preventative measure, administered before surgeries to help reduce the risk of infections. This proactive approach aims to enhance patient outcomes and promote faster recoveries.

A story worth sharing is that of a surgical team who decided to implement a protocol involving penicillin for their orthopedic surgeries. They noticed a marked decrease in postoperative infections compared to previous years, leading to shorter hospital stays for patients and happier outcomes overall. This practice reflects how penicillin continues to adapt to meet the needs of modern healthcare.

Furthermore, the discovery of penicillin derivatives—like amoxicillin—has opened doors to even broader applications. These derivatives maintain the original penicillin's antibiotic properties while enhancing their ability to combat a wider range of bacteria. As a result, doctors can tailor treatments to specific infections more effectively.

Consider Lily, a young mother whose toddler caught a nasty ear infection. After a thorough evaluation, her pediatrician prescribed amoxicillin, knowing it would effectively tackle the bacteria causing the infection while minimizing the chances of side effects. Within days, Lily's little one was back to playing and smiling, demonstrating how penicillin derivatives can offer targeted and effective treatment options in pediatric care.

As we continue to learn more about the applications and combinations of penicillin, it's clear that this antibiotic is far from outdated. The stories of patients like Emma, Mark, and Lily remind us that penicillin's legacy is alive and well in the modern medical landscape. Whether it's through innovative combinations, new applications in chronic conditions, or preventative measures in surgery, penicillin remains a crucial tool in our fight against infections. Its adaptability and the ongoing research surrounding it promise a bright future, ensuring that this storied antibiotic will continue to be a source of healing and hope for generations to come.

6.2 Penicillin in Veterinary Medicine

When we think about penicillin, we often envision its life-saving effects on humans, treating infections that once spelled trouble for our health. However, penicillin has also made significant strides in the world of veterinary medicine, helping our animal companions live healthier lives. Understanding this journey not only sheds light on the versatility of penicillin but also reveals the bond between humans and animals as they share similar health challenges.

Let's take a stroll back in time to the early 20th century when penicillin first burst onto the scene. Its discovery in 1928 by Alexander Fleming was nothing short of revolutionary. As its applications expanded in human medicine, veterinarians began to explore its potential for treating animals. It didn't take long for the veterinary community to recognize that many of the infections that plagued humans also affected pets, livestock, and wildlife.

Consider the story of Buster, a lovable golden retriever who, like many dogs, loved to run and play outdoors. One sunny afternoon, Buster managed to cut his paw on a sharp rock. What seemed like a minor injury quickly escalated into an infection. His worried owner, Sarah, took him to the veterinarian, who diagnosed him with a bacterial infection. Thanks to penicillin, Buster was back on his feet within a week, romping around the yard and chasing squirrels. This simple anecdote highlights the crucial role penicillin plays in treating infections in our furry friends, saving them from suffering and ensuring they can enjoy life to the fullest.

Penicillin is widely used in veterinary medicine to combat various bacterial infections. From common ailments like skin infections and respiratory issues in pets to more serious conditions in livestock, it has proven invaluable. Imagine a farmer with a herd of cattle experiencing an outbreak of mastitis, a painful infection of the udder. Prompt treatment with penicillin can not only relieve the discomfort of the cows but also protect the farmer's livelihood by ensuring a healthy milk supply.

However, administering penicillin in veterinary medicine is not without its challenges. Dosages can vary significantly between species, and veterinarians must be careful to avoid complications. This consideration is especially critical when treating exotic animals or those with unique health needs. For instance, a veterinarian treating a horse with an infection must take into account the animal's size, weight, and overall health to determine the appropriate dosage.

In recent years, veterinarians have also turned to penicillin in more innovative ways. For example, its use in surgeries—whether routine spaying and neutering or more complex procedures—has become standard practice. Preoperative antibiotics, including penicillin, help prevent infections, ensuring a smoother recovery for pets. A personal story comes to mind of a local veterinarian who performed a challenging surgery on a stray cat that had been hit by a car. Administering penicillin before and after the surgery not only safeguarded the cat against infections but also contributed to a remarkable recovery. The cat, now named Whiskers, found a loving home after her ordeal, thanks to the compassionate care she received.

While penicillin is a powerful tool, the veterinary field is also aware of the importance of responsible antibiotic use. Just as in human medicine, the rise of antibiotic resistance is a growing concern. Veterinarians emphasize the importance of using penicillin judiciously, often opting for it only when absolutely necessary. This approach helps preserve its effectiveness, ensuring that it remains a viable option for treating infections in animals for years to come.

Additionally, veterinary medicine continues to evolve alongside human medicine. As research expands, there's a push for new formulations and alternatives to penicillin that can treat infections while minimizing the risk of resistance. One innovative solution is the development of vaccines to prevent bacterial infections in livestock. By reducing the need for antibiotics like penicillin, these vaccines contribute to better animal health and welfare while also addressing public health concerns related to antibiotic use.

The story of penicillin in veterinary medicine is a testament to the intricate bond between humans and animals. Whether it's a beloved pet, a herd of sheep, or a cherished wildlife species, penicillin has played a crucial role in ensuring the health and well-being of our animal companions. From Buster's paw injury to Whiskers' recovery, these anecdotes remind us that the applications of penicillin extend far beyond the human realm. As we continue to explore its potential in veterinary medicine, we celebrate the healing power of this remarkable antibiotic, forging a healthier future for all living beings.

6.3 Alternatives When Penicillin Is Not Effective

While penicillin has undoubtedly been a game changer in the world of medicine, it's not a one-size-fits-all solution. As we've learned, some infections can be stubborn, and there are instances when penicillin simply doesn't do the trick. This reality has led to the development of alternative antibiotics and treatments that play a crucial role in combating infections when penicillin falls short. Understanding these alternatives helps paint a fuller picture of how we can tackle bacterial infections effectively.

Let's think about a time when a beloved family member, perhaps Grandma, got sick. She had been battling a persistent cough and fever that just wouldn't go away. After visiting the doctor, she was prescribed penicillin, but days passed, and her symptoms didn't improve. It turned out that her infection was caused by a strain of bacteria resistant to penicillin, leaving the doctor to consider other options. This is a scenario many of us might be familiar with, where the initial treatment doesn't yield the expected results, prompting a deeper investigation into alternative therapies.

One common alternative is a class of antibiotics known as cephalosporins. Like penicillin, these antibiotics work by interfering with the bacteria's cell wall, but they're often effective against a broader range of bacteria. For example, if Grandma's infection was due to a more resistant bacteria strain, her doctor might switch to a cephalosporin to target it more effectively. This class of antibiotics has been a lifesaver for many patients in similar situations.

Another alternative comes in the form of macrolides, which include antibiotics like azithromycin and erythromycin. These are often used to treat respiratory infections and have been particularly helpful when patients have penicillin allergies. Just imagine the relief for someone who can't take penicillin but needs treatment for a bad case of pneumonia. Macrolides can swoop in and provide effective relief, showing just how important it is to have options in the antibiotic arsenal.

Of course, alternatives don't stop at antibiotics. In some cases, doctors might recommend a combination of antibiotics to cover multiple potential bacterial culprits. This approach is particularly common in serious infections, where the risk of a resistant strain is higher. For instance, if Grandma's infection required a more aggressive treatment plan, her doctor might opt for a combination therapy that blends different antibiotics to effectively tackle the problem.

In recent years, researchers have also been looking into newer classes of antibiotics, such as fluoroquinolones and oxazolidinones, which can treat infections that don't respond to traditional antibiotics. These innovations remind us that the fight against resistant bacteria is ongoing. The development of these new antibiotics often stems from a desire to combat specific resistant strains that have emerged, ensuring that healthcare providers have the tools they need to provide effective care.

But what about non-antibiotic alternatives? This is where things get even more interesting. In some cases, treatment might involve supporting the body's own immune response rather than relying solely on antibiotics. For example, if someone has a mild infection, doctors might recommend rest, hydration, and over-the-counter medications to alleviate symptoms while the body fights off the infection. This supportive care can sometimes be just as important as direct treatment, emphasizing the need for a holistic approach to health.

Consider the story of a young man named Alex who developed a urinary tract infection (UTI). After his doctor prescribed penicillin, Alex found himself feeling worse rather than better. It turned out that his UTI was caused by a strain of bacteria resistant to the medication. Thankfully, his doctor quickly switched gears, prescribing a different antibiotic more suited to tackle the specific strain. Alex was back to his normal self in no time, a testament to the importance of having alternatives in the medical toolbox.

In the realm of veterinary medicine, the need for alternatives is equally significant. Just like with humans, pets can develop infections that penicillin can't treat effectively. In these cases, veterinarians might turn to other antibiotics or even consider treatments like antiseptics and topical medications. A friend of mine had a dog who suffered from a stubborn skin infection that didn't respond to penicillin. The vet recommended a combination of topical treatments and a different class of antibiotics, and soon the dog was healthy and happy again.

The journey of finding the right treatment can sometimes feel like navigating a winding road. When penicillin isn't effective, it's essential to keep exploring alternatives that can provide the needed relief. With advances in medicine and a deeper understanding of bacterial behavior, the future holds promise for new treatments and combinations that will continue to fight infections effectively. It's a reminder of the resilience of both our bodies and the medical community as they work tirelessly to ensure health and well-being for everyone.

Chapter 7

Production and Availability

When we think about penicillin, we often focus on its life-saving properties and its revolutionary impact on medicine. But behind the scenes, there's an intricate process of production and availability that ensures this antibiotic reaches those in need. Understanding this journey not only sheds light on the complexities of pharmaceutical manufacturing but also reminds us of the dedication and innovation that have made penicillin accessible to millions.

Let's take a trip back to the late 1940s, a time when penicillin was just beginning to transform healthcare. Imagine the excitement in hospitals as doctors administered this new wonder drug to patients suffering from severe bacterial infections. However, the challenge was immense. The demand for penicillin skyrocketed, especially during World War II, when soldiers were falling ill from infections that were once fatal. There were stories of doctors racing against time to treat wounded soldiers, relying heavily on this antibiotic to save lives. Yet, as effective as penicillin was, there simply wasn't enough to go around.

The solution came from an unexpected place: the humble mold. Penicillin is derived from *Penicillium* fungi, and scientists discovered that cultivating this mold in large quantities could yield the antibiotic we desperately needed. This led to the establishment of large-scale fermentation processes, where massive tanks were filled with the mold and carefully monitored to optimize penicillin production. It's fascinating to think about how a simple fungus could have such a monumental impact on global health, isn't it?

Fast forward to today, and the production of penicillin has evolved significantly. Modern techniques involve complex biotechnological processes, including genetic engineering and advanced fermentation technology. The aim is not just to produce penicillin efficiently but to ensure that it meets strict safety and quality standards. For instance, many manufacturers employ automated systems that continuously monitor the fermentation environment, adjusting factors like temperature and pH to maximize yield. This meticulous attention to detail is crucial, considering that even minor variations can affect the potency of the final product.

As production methods have advanced, so has the global distribution of penicillin. Today, it's manufactured in various countries around the world, allowing for a more efficient supply chain. However, challenges remain. In some parts of the world, access to penicillin can be limited due to economic or logistical barriers. Consider the story of Maria, a mother living in a rural village. When her young son developed an infection, she rushed to the local clinic only to find that penicillin was out of stock. It's a heart-wrenching scenario that highlights the disparities in healthcare access, even in the age of modern medicine.

To combat these challenges, initiatives are underway to improve the availability of antibiotics in underserved regions. Nonprofit organizations and global health agencies are working tirelessly to ensure that essential medications like penicillin reach those who need them most. For example, outreach programs focus on educating healthcare workers about the importance of proper antibiotic use and ensuring that facilities have the necessary supplies. These efforts underscore the commitment to making penicillin accessible to all, regardless of their location.

But the journey of penicillin production doesn't stop with the manufacturing process. There's a whole other layer when it comes to ensuring its affordability. The cost of antibiotics can vary widely, influenced by factors such as production methods, distribution logistics, and local healthcare systems. Some people might find themselves paying a premium for penicillin, while others may have it available at little to no cost through government programs or charity organizations. This disparity can be frustrating, as it often comes down to where someone lives and their economic circumstances.

Additionally, the rise of antibiotic resistance has spurred renewed interest in the production of penicillin. As bacteria evolve and become resistant to common antibiotics, there's a growing demand for effective treatments. This has led to increased investments in research and development, exploring ways to enhance the efficacy of existing antibiotics and discover new ones. It's a race against time to stay one step ahead of resistant strains, and the production of penicillin is a crucial part of that battle.

Let's not forget about the role of regulations and policies in ensuring the safe production and distribution of penicillin. Governments and health organizations establish guidelines that manufacturers must adhere to, ensuring that antibiotics meet high safety standards. For example, stringent testing is required to confirm the potency and purity of each batch of penicillin before it's released to the market. This meticulous process helps safeguard public health, reassuring patients that the medications they receive are safe and effective.

In a world where healthcare is constantly evolving, the production and availability of penicillin remind us of the importance of innovation, accessibility, and responsibility in medicine. Whether it's through advances in manufacturing technology or global outreach efforts, the commitment to ensuring that this life-saving antibiotic reaches those in need is a testament to our collective dedication to health and well-being. So, the next time you hear about penicillin, remember the incredible journey it takes from the mold in a lab to the hands of patients around the world—each dose carrying with it the promise of healing and hope.

7.1 How Penicillin Is Produced Today

When you think of penicillin, it's easy to imagine a small vial filled with a life-saving liquid, but the journey to that vial is a fascinating story of science, innovation, and a touch of nature's magic. Today, the production of penicillin is a marvel of modern biotechnology, evolving significantly since its serendipitous discovery in the 1920s by Alexander Fleming.

To understand how penicillin is produced today, let's take a closer look at the process, starting with the star of the show: the *Penicillium* mold. You might picture a science lab bustling with activity, and you'd be right. In laboratories around the world, scientists cultivate various strains of this mold, often using advanced techniques to select the most effective ones. Each strain is like a unique artist, producing penicillin in varying quantities and qualities.

Once the right strain is identified, it's time to grow the mold. Imagine a gigantic brewing operation, where large fermenters—think of them as oversized glass jars—are filled with a nutrient-rich broth. This broth is specially designed to provide the mold with everything it needs to thrive. Scientists carefully monitor temperature, pH levels, and oxygen supply, making real-time adjustments to create the perfect environment. It's a bit like being a chef in a kitchen, balancing flavors to achieve a delicious dish, but here the goal is to maximize penicillin production.

As the *Penicillium* grows, it begins to produce penicillin, secreting it into the surrounding broth. This phase can take several days, and it's essential to keep a close eye on the fermentation process. Too little or too much of something can affect the yield. In a story from a local pharmaceutical plant, one technician recalled a batch that yielded half of what they expected because a tiny error in the nutrient mix led to the mold becoming sluggish. It's a reminder that precision in production is crucial, especially when lives are at stake.

Once the fermentation is complete, the next step is harvesting the penicillin. The broth is filtered to separate the mold from the liquid, leaving behind a concentrated penicillin solution. This step is often likened to a winemaker pressing grapes—extracting the essence while leaving behind the rest. The liquid then undergoes further purification processes, including crystallization, which transforms it into a stable form that can be more easily transported and stored.

Interestingly, penicillin isn't just produced in one form. There are different types of penicillin, each with its specific uses. For instance, some are formulated for oral consumption in tablets, while others are designed for injections in hospitals. The production process adapts to these different needs, ensuring that each form retains its potency and effectiveness.

Quality control is paramount in today's production of penicillin. Manufacturers adhere to stringent regulations set by health authorities to ensure that every batch meets safety and efficacy standards. Before penicillin is released to the market, it undergoes rigorous testing, including checking for purity, potency, and potential contaminants. A story shared by a quality control manager highlighted a time when a batch had to be discarded due to a minor contamination issue. Although it was disappointing, the commitment to safety ensures that patients receive only the best.

In recent years, the production of penicillin has also embraced technology. Automation and artificial intelligence are now playing a significant role in monitoring fermentation processes, predicting optimal conditions for mold growth, and even improving the efficiency of purification methods. These advancements allow for faster production times and a more consistent product. It's like having a smart assistant in the lab, constantly analyzing data and suggesting improvements.

Furthermore, there's a growing emphasis on sustainability in the production of antibiotics, including penicillin. Manufacturers are exploring environmentally friendly practices, such as reducing waste and minimizing energy consumption. For instance, some companies are investing in bioreactors that recycle resources, creating a more sustainable cycle of production.

The journey of penicillin from the laboratory to the pharmacy shelf is a testament to human ingenuity and the power of nature. With each vial that is filled, there's a story of dedication, precision, and a commitment to improving health. It's a reminder that behind every dose of penicillin lies the hard work of scientists, technicians, and countless others, all striving to make the world a healthier place. As we look to the future, the ongoing innovations in penicillin production promise not only to preserve its legacy but to enhance its role in modern medicine, ensuring that it remains a cornerstone of healthcare for generations to come.

7.2 Challenges in Supply and Distribution

The journey of penicillin from a lab bench to the hands of a patient is filled with hurdles, especially when it comes to supply and distribution. While we often celebrate the life-saving properties of antibiotics, it's essential to recognize the behind-the-scenes challenges that can make access to these medications a struggle for many.

Imagine a bustling clinic in a rural area of a developing country. Nurses and doctors are doing their best to treat patients, but their shelves are often empty. This situation is not uncommon. Many healthcare facilities face significant challenges in obtaining the medications they need, and penicillin is no exception. The distribution of antibiotics can be influenced by a myriad of factors, from economic constraints to logistical issues.

One major hurdle is the lack of infrastructure. In many remote regions, roads may be poorly maintained, making it difficult for suppliers to reach clinics. A poignant example comes from a health worker named Paul who serves in a mountainous region. He recalls a time when he had to transport supplies on foot for hours, navigating treacherous paths to deliver penicillin to a small village that had run out. His commitment to the community was unwavering, but he knew that this kind of delivery system couldn't be relied upon long-term.

Supply chain disruptions are another significant concern. Natural disasters, political instability, or even global events like the COVID-19 pandemic can affect the availability of penicillin. During the pandemic, many healthcare systems around the world faced unprecedented challenges. Hospitals were overwhelmed, and the demand for antibiotics surged. In some cases, production facilities had to halt operations, leading to a ripple effect that left clinics without essential medications. This is particularly concerning because it can delay treatment for infections, leading to complications for patients.

Economic factors also play a significant role in the supply of penicillin. Many developing countries struggle with limited budgets for healthcare, which can impact the procurement of medications. When funds are tight, essential supplies often get prioritized over antibiotics. In these situations, healthcare professionals have to make tough choices about how to allocate limited resources, sometimes leaving patients vulnerable to untreated infections.

The issue of counterfeit medications further complicates the landscape. In regions where access to genuine penicillin is limited, some unscrupulous suppliers may fill the gap with fake products. These counterfeit drugs can be ineffective or even harmful, putting patients at risk. A nurse named Fatima shared her concern about a patient who received a counterfeit antibiotic that did not help their infection. Fortunately, she was able to intervene, but many patients may not have the same luck. Raising awareness about the importance of sourcing medications from reputable suppliers is crucial to combat this issue.

Efforts to tackle these supply and distribution challenges are underway. Collaborations between governments, international organizations, and non-profits aim to improve healthcare delivery systems. For instance, initiatives to enhance infrastructure—like building better roads or providing reliable transportation—can make a difference in getting medications where they need to go.

Furthermore, technology is playing an increasingly vital role in addressing these challenges. In some countries, mobile health apps are being used to track inventory and manage supplies more efficiently. These innovations help healthcare providers ensure that essential medications like penicillin are available when needed. In rural areas, community health workers equipped with tablets can report stock levels and request supplies directly, improving communication with suppliers and streamlining the distribution process.

Despite these positive strides, the journey to ensuring consistent access to penicillin remains fraught with challenges. Personal stories of healthcare workers and patients highlight the urgency of addressing these issues. A grandmother who lost her grandchild to an infection because of a lack of antibiotics serves as a somber reminder of what's at stake. Her experience underscores the need for a robust supply chain and effective distribution strategies to ensure that no one suffers due to preventable illnesses.

While penicillin is a remarkable tool in modern medicine, its journey to patients is complex. By recognizing and addressing the challenges in supply and distribution, we can work toward a future where everyone, regardless of where they live, has access to this life-saving antibiotic. It's a collective effort that requires commitment from all sectors—government, healthcare, and community—to ensure that penicillin is available when it matters most.

7.3 Access to Penicillin in Developing Countries

Imagine a small village in a developing country where a child wakes up with a high fever, a sign of a possible bacterial infection. In many parts of the world, access to antibiotics like penicillin can be a matter of life and death, but for this child, the situation is complex. The story of penicillin in developing countries is not just about a medication; it's a tale woven with threads of hope, challenges, and the tireless efforts of communities and healthcare workers.

Penicillin, discovered in 1928, revolutionized medicine and has saved countless lives. However, despite its critical importance, access to this antibiotic remains inconsistent in many developing countries. A 2018 report from the World Health Organization highlighted that while penicillin is essential for treating infections, availability can be patchy, especially in rural areas. Sometimes, health clinics might run out of supplies or lack the infrastructure to store medications properly. This can leave families in desperate situations, knowing that effective treatment exists but is just out of reach.

In some cases, people have to travel long distances to access healthcare facilities that have penicillin. I remember hearing about a woman named Maria from a rural community in Central America. When her young son fell ill, she walked for miles to reach the nearest clinic. When she finally arrived, she was relieved to find that penicillin was available, but the experience left her exhausted and worried about how she would afford transportation in the future. Maria's story is just one of many, illustrating the lengths people will go to seek medical care.

The disparities in access to penicillin often come down to systemic issues, including poverty and inadequate healthcare infrastructure. In some developing countries, healthcare systems are strained, with limited resources for training healthcare professionals and ensuring the proper distribution of medicines. A dedicated nurse in a small clinic might be managing hundreds of patients a week, struggling to keep track of medications and supplies. They do their best, but without adequate support, it becomes challenging to provide optimal care.

Efforts to improve access to penicillin are underway in various parts of the world. Organizations like Médecins Sans Frontières (Doctors Without Borders) work tirelessly to provide essential medicines, including penicillin, to communities in need. They often set up mobile clinics and reach out to remote areas, offering not just medications but education on the importance of antibiotics and proper treatment practices. Their impact is profound, providing hope and relief to families who might otherwise be left to fend for themselves.

Partnerships between governments, international organizations, and local communities are also crucial in addressing access issues. Initiatives that focus on strengthening healthcare systems, improving supply chains, and training healthcare workers can significantly enhance the availability of penicillin. For instance, in countries like Kenya and Tanzania, community health programs have been established to educate villagers about common infections and the importance of seeking timely treatment. These programs empower communities, giving them the knowledge they need to advocate for better healthcare services.

Despite the progress being made, challenges remain. In some regions, the rising issue of antibiotic resistance complicates treatment options. Inadequate access to penicillin can lead to improper usage, with people using leftover medications or self-medicating without professional guidance. This can create a cycle of resistance that jeopardizes the effectiveness of antibiotics. The story of penicillin is not just about saving lives; it's also about ensuring that it continues to be a reliable treatment for generations to come.

Personal stories, like that of a young man named Samuel who was treated for pneumonia with penicillin, bring this issue to life. After receiving timely treatment,

Samuel was able to return to his studies, inspiring others in his community to seek medical help when needed. His journey illustrates the ripple effect of access to essential medications—not only does it save lives, but it also enables individuals to pursue their dreams and contribute positively to society.

As we reflect on the state of penicillin access in developing countries, it's essential to remember that every effort counts. The work being done by healthcare workers, community organizations, and advocates is critical in bridging the gaps. The hope is that one day, access to life-saving antibiotics like penicillin will be universal, ensuring that no child has to suffer because of a lack of medicine. It's a goal worth striving for, and together, we can create a future where everyone has the opportunity to thrive, free from the burden of preventable infections.

Chapter 8

Controversies and Ethical Considerations

As we dive into the world of penicillin, it's essential to understand that its story is not just about medical breakthroughs and life-saving properties. It also encompasses a complex tapestry of controversies and ethical considerations that have shaped its use and impact on society. From the early days of its discovery to the challenges we face today, the conversation around penicillin often sparks heated debates, ethical dilemmas, and moral questions that deserve our attention.

Consider the early years after Alexander Fleming discovered penicillin in 1928. The excitement surrounding this miraculous drug was palpable. It was hailed as a game changer in medicine, capable of treating previously deadly infections. Yet, this initial enthusiasm was soon tempered by ethical concerns over how it would be produced and distributed. In the 1940s, during World War II, the race to produce penicillin ramped up. Factories were hastily built, and production processes were refined. But as the demand skyrocketed, the question emerged: Who would get access to this precious medicine? Soldiers? Civilians? Those with more means? These questions reveal the complicated landscape of equitable access to healthcare.

Fast forward to today, and the ethical considerations surrounding penicillin continue to evolve. One major concern is the issue of antibiotic resistance. As penicillin has been used widely, some bacteria have developed the ability to resist its effects. This resistance raises serious questions about our reliance on antibiotics and how we prescribe them. A story that illustrates this dilemma is that of a young mother named Sarah. After her son was prescribed antibiotics for an ear infection, she was diligent about completing the course. However, she later learned that overprescribing antibiotics can contribute to resistance. This revelation left her feeling conflicted about her role in this larger public health issue.

The question of whether to prescribe penicillin, especially in the face of resistance, can create ethical quandaries for healthcare providers. Should they prescribe antibiotics to alleviate symptoms, knowing that it may contribute to a bigger problem? Or should they withhold treatment to avoid exacerbating resistance? These decisions can weigh heavily on doctors, who are often faced with time constraints and patient pressure.

Another controversy revolves around the use of penicillin in agriculture. Many farmers have turned to antibiotics like penicillin to promote growth in livestock and prevent disease in herds. While this practice can enhance productivity, it raises concerns about the long-term impact on public health. The use of antibiotics in animals can lead to the development of resistant strains of bacteria, which can then enter the human food chain. For example, a farmer named

Joe shared his experience of transitioning to organic practices, expressing his desire to protect his family's health and contribute to sustainable agriculture. He faced criticism from peers who relied on antibiotics for quick profits, highlighting the ethical dilemma of balancing economic interests with health considerations.

Then there's the issue of access to penicillin in developing countries. Despite its low cost and effectiveness, many people in low-income areas struggle to obtain this vital medication. A powerful anecdote comes from a healthcare worker named Maria, who works in a rural clinic. She often encounters patients who cannot afford basic antibiotics, leading to preventable deaths. The frustration of knowing that a simple injection could save lives but remains out of reach for many fuels her passion for advocating for equitable healthcare. Maria's story highlights the ethical obligation we have to ensure that life-saving medications are available to all, regardless of their socio-economic status.

Additionally, the conversation around ethical sourcing of penicillin is becoming increasingly important. In an era where transparency and corporate responsibility are in the spotlight, consumers are demanding to know more about where and how their medications are produced. A growing number of individuals want to support companies that adhere to ethical practices, from fair labor conditions to sustainable sourcing. This shift in consumer behavior is prompting pharmaceutical companies to reevaluate their practices and consider the ethical implications of their supply chains.

The journey of penicillin is not just about its medical marvels; it is also a reflection of the ethical considerations that come with it. As we navigate the complexities of antibiotic resistance, equitable access, agricultural practices, and sourcing, it's crucial to engage in these discussions with empathy and awareness. By sharing stories like those of Sarah, Joe, and Maria, we humanize the debate, bringing to light the real people affected by these issues. As we look to the future, we must consider not only the efficacy of penicillin but also the ethical implications of its use, ensuring that this life-saving medication remains a beacon of hope for all.

8.1 Overuse of Antibiotics and the Consequences

In a world where antibiotics have become household names, the phrase "more is better" often echoes in our minds. We want to ensure that we tackle every infection and illness that comes our way, especially when it seems like a simple pill could do the trick. However, this mindset has led to a significant problem: the overuse of antibiotics.

Take a moment to think about the story of Lisa, a young mother who rushed her son, Jake, to the doctor after he developed a bad cough. The doctor quickly diagnosed it as a viral infection, which antibiotics cannot treat. Despite this, Lisa felt uneasy about not receiving a prescription. "What if it gets worse?" she wondered. The doctor, understanding her concerns, reassured her that most viral infections resolve on their own and that antibiotics would do more harm than good. Lisa left the office feeling conflicted. Unfortunately, her story is not unique. Many patients like Lisa often leave the doctor's office feeling pressured to receive antibiotics, even when they aren't necessary.

This pattern of prescribing antibiotics for viral infections, like colds or the flu, contributes significantly to the problem of antibiotic overuse. Studies show that nearly a third of antibiotics prescribed in outpatient settings are unnecessary. With a growing list of bacterial infections, the consequences of this over-prescription can be dire. The most alarming outcome is the rise of antibiotic resistance, a phenomenon that transforms once-treatable infections into potential killers.

Think back to the time when penicillin was first introduced. It was hailed as a miracle drug, revolutionizing the treatment of bacterial infections. However, over the decades, its indiscriminate use has allowed bacteria to adapt and evolve. A well-known example is methicillin-resistant Staphylococcus aureus, or MRSA, a strain of bacteria that has become resistant to many antibiotics, including penicillin. This can be especially frightening in hospital settings, where vulnerable patients are at a higher risk of infections that can no longer be easily treated.

Moreover, antibiotic overuse doesn't just affect individual patients. It poses a threat to public health as a whole. The more we use antibiotics, the more we allow bacteria to evolve and become resistant. This cycle creates a scenario where common infections can become life-threatening. Imagine a world where minor surgeries become risky because of the fear of postoperative infections that can't be controlled with antibiotics.

Let's not forget the role of agriculture in this conversation. In many countries, antibiotics are used not only to treat sick animals but also to promote growth in healthy livestock. This practice raises ethical questions about food production and public health. For instance, a farmer named Tom shared his journey of transitioning to antibiotic-free farming. Initially met with skepticism from fellow farmers who relied on antibiotics for quick profits, Tom's decision was rooted in a commitment to both animal welfare and consumer health. By choosing not to use antibiotics in his livestock, he hopes to contribute to the broader effort of reducing antibiotic resistance in both humans and animals.

The consequences of antibiotic overuse extend beyond the medical community; they touch every corner of society. In many developing countries, access to antibiotics is often unregulated, leading to self-medication and misuse. Stories abound of individuals taking leftover antibiotics from a previous illness or purchasing them without a prescription. These practices not only contribute to resistance but also put individuals at risk of serious health complications.

So, what can we do to combat this growing crisis? Education is a crucial first step. Patients need to understand that antibiotics are not a cure-all and that overusing them can lead to serious consequences. Healthcare providers play a vital role in this, ensuring that patients are informed and feel supported in their decisions. Encouraging open conversations about treatment options can help build trust and alleviate the pressure many patients feel.

While antibiotics like penicillin have undeniably changed the landscape of medicine for the better, their overuse poses a real threat to public health. The stories of individuals like Lisa and Tom highlight the personal and collective stakes involved. As we navigate this complex issue, it is essential to prioritize education, open dialogue, and responsible use of antibiotics to ensure that future generations can continue to benefit from these remarkable drugs without falling prey to the consequences of overuse.

8.2 Ethical Issues in Antibiotic Distribution

Imagine standing in a bustling pharmacy, where the air is filled with the faint scent of antiseptic and the chatter of customers picking up their prescriptions. Among the shelves stocked with various medications, antibiotics sit prominently, promising relief from infections and illness. But behind this seemingly simple scene lies a complex web of ethical issues surrounding the distribution of these critical drugs.

One story that often comes to mind is that of Sarah, a dedicated nurse working in a rural clinic. She has seen firsthand the struggles her patients face, not just with illnesses but with accessing the treatments they need. One day, a young mother brought in her feverish child, desperately hoping for a prescription that would ease her child's suffering. Sarah examined the child and determined that an antibiotic was indeed necessary. However, due to a shortage in their clinic's supply, they didn't have the medication on hand. This moment highlighted a crucial ethical dilemma: how do healthcare providers decide who gets access to essential medications when supply is limited?

In many regions, especially in developing countries, the distribution of antibiotics is fraught with challenges. These can range from supply chain issues to pricing disparities. There's a heartbreaking irony here. While some people in wealthier nations may have an abundance of antibiotics at their fingertips, others across the globe are struggling to secure even the most basic treatments. This disparity raises questions about equity in healthcare and the moral responsibilities of pharmaceutical companies and governments.

Take the story of Ahmed, a father in a rural village in Africa. When his daughter fell ill with a bacterial infection, he traveled miles to the nearest pharmacy, only to find that the antibiotics he needed were either out of stock or priced beyond his means. Feeling helpless, he resorted to seeking alternative treatments that might not be safe or effective. In Ahmed's case, the ethical issue of accessibility becomes painfully clear. Shouldn't everyone, regardless of their location or financial situation, have access to lifesaving medications?

Moreover, the ethical considerations don't stop at accessibility. The quality of antibiotics available also matters significantly. Counterfeit or substandard medications are a growing problem in many parts of the world. For instance, a recent investigation uncovered that nearly one in ten medicines sold in developing countries is either fake or poorly manufactured. Such issues not only jeopardize patient health but also undermine trust in healthcare systems. Imagine the despair of a patient who, after struggling to obtain antibiotics, discovers that the medication they received might not work at all. This scenario highlights the need for robust regulations and oversight in the distribution of antibiotics to protect patients and ensure their safety.

The pharmaceutical industry itself grapples with ethical challenges related to antibiotic distribution. Profit motives can often overshadow the urgent need for accessible, affordable medications. A striking example is the price of certain antibiotics that have become scarce due to dwindling interest from pharmaceutical companies. As some manufacturers cut back on production to focus on more profitable drugs, patients are left vulnerable. The ethical question arises: should profit take precedence over patient care?

We can't overlook the responsibility of healthcare providers and policymakers, either. They play a crucial role in ensuring that antibiotics are used responsibly and that patients receive the right medications when needed. The misuse and over-prescription of antibiotics have long-lasting consequences, leading to resistance that jeopardizes their effectiveness. This places an ethical burden on healthcare professionals to educate their patients about the appropriate use of these powerful drugs.

In light of these issues, what can be done? A multi-faceted approach is essential. Governments and organizations must work together to improve supply chains, ensuring that antibiotics are available where they are needed most. Furthermore, investments in education can empower patients to make informed decisions about their health and the medications they take. Pharmaceutical companies need to be encouraged to prioritize equitable access over profit margins, fostering a more just healthcare system.

The ethical issues surrounding antibiotic distribution are complex and far-reaching. The stories of individuals like Sarah and Ahmed serve as reminders of the real-world implications of these challenges. As we navigate this landscape, it's vital to keep the focus on empathy and equity, ensuring that antibiotics remain a tool for healing rather than a source of inequality. By addressing these ethical dilemmas, we can work toward a future where every individual, regardless of their background or location, has access to the treatments they need to live healthy lives.

8.3 Balancing Treatment with Prevention of Resistance

In our ongoing battle against infections, antibiotics have become indispensable weapons. Yet, as we rely more on these powerful drugs, we find ourselves in a delicate balancing act: treating illnesses while also preventing antibiotic resistance. This challenge is not just a medical dilemma but a societal one, echoing stories of individuals and communities striving to stay healthy in an increasingly complex world.

Take the case of Emily, a lively ten-year-old who loves playing soccer with her friends. One afternoon, she came home with a high fever and a sore throat. Her concerned parents took her to the doctor, who quickly diagnosed her with strep throat. Fortunately, Emily was prescribed penicillin, which worked wonders. Within a couple of days, she was back on the field, dribbling the ball and enjoying life. However, this happy ending comes with a cautionary tale. Every time antibiotics like penicillin are used, there's a risk that some bacteria may survive and evolve into resistant strains.

Emily's story reminds us that while antibiotics can be lifesaving, they also require responsible usage. Unfortunately, this isn't always the case. Antibiotic overprescription has become a widespread issue, often driven by patients' demands for immediate solutions. Imagine a scenario where a parent goes to the doctor with a child suffering from a viral infection, such as the flu. The parent, anxious for relief, requests antibiotics, believing that they will help. However, antibiotics have no effect on viruses. Yet, many doctors find themselves prescribing antibiotics to appease worried parents, leading to unnecessary use.

To illustrate the impact of this behavior, let's turn to a global perspective. In some countries, antibiotics are available over the counter, meaning people can purchase them without a prescription. While this may seem convenient, it can lead to rampant misuse. People might take antibiotics for minor ailments, or even share them with friends and family, unknowingly contributing to resistance. This practice has real-world consequences: the World Health Organization has warned that antibiotic resistance could lead to millions of deaths worldwide by 2050 if not addressed.

So, how can we balance the need for effective treatment with the prevention of resistance? One promising strategy is antibiotic stewardship. This approach involves promoting the appropriate use of antibiotics, ensuring they are prescribed only when necessary and in the right dosages. For healthcare providers, this means taking the time to educate patients about the difference between bacterial and viral infections. It might also mean being firm in discussions, explaining why antibiotics are not needed in certain situations.

For example, consider Dr. Martinez, a family physician who makes it a point to involve her patients in the decision-making process. When a worried parent comes in with a child suffering from a cold, she takes a moment to explain how antibiotics won't help and emphasizes the importance of rest, hydration, and time. Dr. Martinez has noticed that this approach not only eases parents' fears but also fosters a better understanding of antibiotic use.

Education plays a crucial role here, not just for patients but for the general public. Awareness campaigns highlighting the risks of misuse can go a long way in changing perceptions. Just think of a community health initiative where local schools, clinics, and pharmacies collaborate to share information about antibiotics. These efforts can empower families to make informed choices and promote healthier habits, such as using proper hand hygiene to prevent infections in the first place.

Moreover, advancements in technology and research are providing new avenues for addressing antibiotic resistance. Innovative approaches, such as developing rapid diagnostic tests, can help determine the cause of an infection more quickly, allowing for appropriate treatment without unnecessary antibiotic use. Imagine a future where doctors can confidently diagnose a patient in a matter of minutes, pinpointing whether a bacterial infection is present and what specific antibiotic would be most effective.

As we navigate this intricate landscape, it's essential to remember the stories behind the statistics. Each case of resistance has a human face, and behind every antibiotic prescription lies a narrative. The balance between treatment and resistance is not just a medical issue; it's about the lives impacted by our choices. For every Emily who recovers quickly, there may be another child facing the consequences of resistance in the future.

In the end, it's about cultivating a culture of responsibility and understanding. By prioritizing education, fostering collaboration among healthcare providers, and promoting stewardship practices, we can work toward a future where antibiotics remain effective tools for healing, allowing children to enjoy their soccer games without the looming threat of resistant infections. Together, we can make informed decisions that not only treat today's ailments but also safeguard the health of future generations.

Chapter 9

The Future of Penicillin

As we step into a new era of medicine, the future of penicillin holds a fascinating mix of hope and challenge. This humble antibiotic, discovered nearly a century ago by Alexander Fleming, has changed the landscape of healthcare and continues to save countless lives. Yet, the evolution of bacteria and the emergence of antibiotic resistance pose significant hurdles that we must navigate together.

Imagine a world where a simple infection, once easily treatable with penicillin, could become a life-threatening condition due to resistance. This is not just a hypothetical scenario; it's a reality that researchers, healthcare professionals, and policymakers are grappling with today. As we reflect on the legacy of penicillin, we must also consider what lies ahead and how we can harness innovation to ensure its continued effectiveness.

Let's take a moment to look back at how penicillin changed the course of medicine. During World War II, penicillin was hailed as a miracle drug that drastically reduced the death toll from infections. Soldiers who would have succumbed to wounds became survivors, returning home to their families. This story isn't just about a medication; it's about the lives transformed and the families united. My grandmother often told me stories of how penicillin saved my great-uncle after he was injured in the war. For many, that miracle became a lifeline, demonstrating the profound impact of antibiotics on human health.

However, as we celebrate these successes, we also face a sobering reality. The overuse and misuse of antibiotics have led to the rise of resistant bacteria, raising questions about the sustainability of our current practices. To address this issue, scientists and researchers are tirelessly working to develop new strategies and solutions. The future of penicillin might not only be about creating new antibiotics but also about improving how we use existing ones.

One exciting avenue being explored is the development of synthetic antibiotics. Researchers are studying ways to create antibiotics that mimic the structure and function of penicillin but with enhanced properties to overcome resistance. Imagine a new class of drugs that could target resistant bacteria while leaving beneficial microbes unharmed. This isn't just wishful thinking; it's actively being researched in laboratories around the globe.

Additionally, there's a growing focus on personalized medicine, where treatments are tailored to individual patients based on their genetic makeup. This could mean that rather than using a one-size-fits-all approach, doctors could prescribe antibiotics that are more effective for specific infections in specific individuals. It's a future where treatments are not only more efficient but also reduce the risk of resistance developing.

To further illustrate the potential of personalized medicine, let's look at the story of Sarah, a

32-year-old woman who suffered from recurrent urinary tract infections (UTIs). After countless rounds of antibiotics, Sarah's doctor decided to take a different approach. Through genetic testing, they identified a specific strain of bacteria that was resistant to many common antibiotics. Armed with this knowledge, they tailored a treatment plan just for Sarah, incorporating a targeted antibiotic that effectively cleared the infection. Sarah's experience highlights how advancements in medical technology could revolutionize the way we approach antibiotic treatment in the future.

Education will also play a critical role in the future of penicillin. As awareness about antibiotic resistance grows, public health initiatives are essential in promoting responsible antibiotic use. Imagine school programs that teach children about the importance of hygiene, vaccination, and the dangers of antibiotic overuse. By instilling these values early on, we can cultivate a generation that understands the significance of antibiotics and their role in public health.

Moreover, collaboration is key. The future of penicillin won't be shaped by any one group alone. Researchers, healthcare providers, policymakers, and patients must come together to create comprehensive strategies to combat resistance and ensure access to effective treatments. Think of it like a team effort; just as soldiers in a battle rely on one another, we too must unite to face the challenges ahead.

However, we must also acknowledge the disparities in access to antibiotics, particularly in low- and middle-income countries. As we explore the future of penicillin, it's vital to advocate for equitable access to medications. Imagine a world where every individual, regardless of their location, can receive life-saving antibiotics. This is not merely a dream; it's a goal that requires commitment and action from the global community.

The future of penicillin is a story still being written. With each new discovery and every innovative approach, we have the opportunity to shape a healthcare landscape where antibiotics remain effective allies in our fight against infections. By combining science, education, collaboration, and a commitment to equity, we can ensure that the legacy of penicillin continues to thrive.

As we move forward, let's carry with us the stories of those whose lives have been saved by this remarkable drug, keeping in mind the responsibility we have to protect its power for future generations. The journey ahead may be challenging, but with collective effort, we can secure a future where penicillin—and antibiotics as a whole—continues to be a beacon of hope in the realm of medicine.

9.1 Innovations in Antibiotic Research

The landscape of antibiotic research is evolving faster than ever, driven by necessity as we face the daunting challenge of antibiotic resistance. But what does the future hold? What innovations are emerging that promise to change how we fight infections? To answer these questions, let's take a journey through some of the most exciting developments in antibiotic research today.

One of the most promising avenues is the exploration of *bacteriophages*, viruses that specifically target and kill bacteria. Picture this: a tiny virus, invisible to the naked eye, invading a bacterial cell and replicating until it bursts, effectively destroying the bacteria from within. This isn't science fiction; it's real science, and researchers are finding that these natural predators can be powerful allies in our fight against resistant strains.

Anecdotes of successful phage therapy treatments remind us of its potential. Take, for example, the story of a young boy named Isaac who suffered from a severe bacterial infection that conventional antibiotics could not touch. After months of treatment with no success, doctors turned to bacteriophages. They tailored a phage cocktail specifically for Isaac's infection, and within days, he began to recover. Isaac's story highlights a beacon of hope for many patients facing the limits of traditional antibiotics.

In addition to bacteriophages, scientists are also delving into the world of *antimicrobial peptides*. These naturally occurring molecules are part of the immune response in many organisms, including humans. They act like miniature warriors, attacking bacteria by disrupting their cell membranes. The beauty of antimicrobial peptides lies in their potential versatility; researchers are experimenting with ways to enhance their effectiveness or modify their structures to make them more potent against resistant bacteria. The excitement surrounding this research stems from the idea that we might be able to develop entirely new classes of antibiotics that work in ways we've never seen before.

Another innovative approach is the use of *AI and machine learning* to discover new antibiotics. Imagine a computer analyzing thousands of compounds and predicting which ones could effectively combat specific bacteria. Researchers are harnessing the power of AI to sift through massive databases of chemical compounds, identifying promising candidates for antibiotic development. This technology not only accelerates the research process but also opens up a world of possibilities that traditional methods might miss.

Consider the story of a research team that used AI to discover a novel antibiotic called *teixobactin*. Traditionally, it was incredibly challenging to cultivate certain bacteria in the lab, limiting the pool of antibiotic candidates. However, by utilizing AI, they were able to identify compounds that had previously gone unnoticed. Teixobactin showed remarkable effectiveness against Gram-positive bacteria, including some resistant strains, making it a significant breakthrough.

Moreover, the field of *combination therapy* is gaining momentum. Researchers are investigating how pairing existing antibiotics with novel compounds can enhance their effectiveness against resistant bacteria. This strategy not only has the potential to overcome resistance but also to extend the lifespan of existing antibiotics. The notion of combining therapies mirrors our everyday lives, where collaboration often leads to better outcomes. By working together, antibiotics can potentially enhance each other's effects, much like a well-coordinated team in a sporting event.

Let's not overlook the role of *vaccine development* in combating antibiotic resistance. Vaccines can prevent infections in the first place, reducing the need for antibiotics. Innovative approaches, such as *mRNA vaccines*, which gained attention during the COVID-19 pandemic, are being explored to develop vaccines against bacterial infections as well. The success of these vaccines may lead to new ways of preventing diseases that previously required antibiotic treatment.

As we discuss these innovations, it's essential to remember the human element behind the research. Scientists, researchers, and healthcare professionals are driven by real-world stories of patients struggling with infections. Their dedication to finding solutions is fueled by a passion to improve lives and a commitment to public health.

The innovations in antibiotic research are a testament to human ingenuity and resilience. From bacteriophages to AI-driven drug discovery, each advancement offers hope in our battle against antibiotic resistance. These stories of innovation are not just about scientific progress; they're about the people who will benefit from these breakthroughs. As we move forward, we must continue to support and invest in this critical research, ensuring that we have the tools we need to keep infections at bay and protect future generations. The future of antibiotics is not just a scientific pursuit; it's a journey towards healthier lives for everyone.

9.2 Penicillin Derivatives and Their Uses

When we think of penicillin, the mind often wanders to its discovery and the revolutionary impact it had on medicine. However, the story doesn't end there. Over the decades, scientists have crafted various penicillin derivatives, each with its unique applications and benefits. These derivatives have not only expanded our arsenal against infections but have also provided a glimpse into the ongoing evolution of antibiotic therapy.

Let's start with a classic example: amoxicillin. This widely used penicillin derivative has become a staple in treating various infections, from earaches to pneumonia. Its versatility is one of its strongest attributes. Amoxicillin can tackle both Gram-positive and certain Gram-negative bacteria, making it a go-to choice for many healthcare providers.

One memorable story involves a little girl named Mia, who faced recurring ear infections. After several visits to the pediatrician, her doctor prescribed amoxicillin. Within days, Mia was back to her usual energetic self, playing in the park and laughing with her friends. Her experience illustrates how a penicillin derivative can transform lives, restoring health and happiness.

Another derivative worth mentioning is *ampicillin*. This broad-spectrum antibiotic is often employed in more serious infections, including those caused by *Listeria monocytogenes*, a bacterium that can lead to meningitis and sepsis, especially in vulnerable populations like pregnant women and the elderly. What sets ampicillin apart is its ability to penetrate the blood-brain barrier, which makes it especially valuable in treating central nervous system infections.

An interesting historical note about ampicillin is that it was initially introduced in the 1960s as a way to combat the growing threat of bacterial resistance. Doctors were concerned about the limitations of penicillin itself, and ampicillin emerged as a more robust option, showcasing how antibiotic development responds to changing medical needs.

In addition to these, there's *oxacillin*, which was specifically designed to combat penicillin-resistant staphylococci. Imagine a surgeon in the operating room, anxious about potential infections post-surgery. The introduction of oxacillin offered a powerful solution to this worry, allowing doctors to tackle infections that were once formidable opponents. In this way, oxacillin not only serves as an antibiotic but also as a tool of reassurance for both patients and healthcare providers.

Then there's *piperacillin*, often combined with another beta-lactamase inhibitor called tazobactam. This combination enhances its effectiveness against a broader range of bacteria, including those that produce enzymes capable of resisting antibiotics. It's particularly useful in treating severe infections in hospitals, where resistant strains are more prevalent. In one case, a patient battling a complicated urinary tract infection found success with this duo after multiple failed treatments. The blend of piperacillin and tazobactam not only cured the infection but also underscored the importance of innovation in antibiotic therapy.

We can't overlook *cloxacillin*, another derivative that finds its niche in treating skin infections, particularly those caused by *Staphylococcus aureus*. This derivative has been invaluable in dermatology, with many patients experiencing rapid improvements in their skin conditions. One particular patient shared how cloxacillin turned around a stubborn skin infection that had been troubling her for months. After just a week of treatment, she felt a sense of relief and regained her confidence.

As we reflect on these derivatives, it's important to note that they are not without challenges. Just like penicillin, these antibiotics can lead to resistance if overused or misused. This reality emphasizes the need for careful prescribing practices and patient education. It's not just about using the right medication; it's about using it responsibly.

In the grand scheme, the derivatives of penicillin illustrate how far we've come in our fight against bacterial infections. Each derivative brings its own set of capabilities, allowing healthcare providers to tailor treatments to individual patient needs. They are more than just drugs; they represent the ingenuity of science and the relentless pursuit of better health outcomes.

The world of penicillin derivatives is rich and complex, filled with stories of innovation, resilience, and transformation. As we continue to navigate the challenges of antibiotic resistance, these derivatives stand as a testament to the advancements in medical science and the hope they bring to countless individuals. Just like Mia, who bounced back from her ear infection, many lives have been touched by these remarkable antibiotics, reminding us of the importance of continued research and responsible use in this ever-evolving field.

9.3 Global Initiatives to Preserve Antibiotic Effectiveness

As we stand at a crossroads in the battle against antibiotic resistance, the global community has rallied together to preserve the effectiveness of these life-saving drugs. The alarming rise of resistant bacteria is not just a local issue; it's a global crisis that affects us all. This realization has spurred initiatives that unite governments, healthcare organizations, and communities worldwide, working together to safeguard antibiotics for future generations.

One of the most significant efforts is spearheaded by the World Health Organization (WHO). In 2015, they launched the Global Action Plan on Antimicrobial

Resistance, urging countries to develop their own national action plans. Countries like Sweden and the Netherlands have been frontrunners in this initiative, successfully reducing antibiotic use while maintaining health standards. In Sweden, for instance, public health campaigns raised awareness about the dangers of unnecessary antibiotic prescriptions, leading to a remarkable decrease in usage without compromising patient care.

Imagine a family doctor in a small Swedish town who carefully assesses each patient's symptoms before deciding on treatment. Through engaging community dialogues and educational programs, patients learn that antibiotics aren't always the answer. This proactive approach not only protects antibiotics but also empowers patients to be part of the solution.

Another notable initiative is the Antibiotic Resistance Coalition, a partnership of multiple organizations dedicated to raising awareness and advocating for change. Their campaigns have taken on various forms, from social media movements to community workshops. They share personal stories that resonate with people, illustrating the real-life impact of antibiotic resistance. For example, a mother in the coalition might share her harrowing experience of losing her child to a drug-resistant infection, sparking discussions on the importance of responsible antibiotic use.

In addition to awareness campaigns, many countries are implementing stringent regulations on antibiotic prescriptions. In India, where antibiotic misuse has reached alarming levels, new policies aim to curb over-the-counter sales of these medications. This initiative encourages pharmacists to ask more questions before dispensing antibiotics, ensuring that patients only receive them when truly necessary. A small, rural clinic in India can become a powerful example of this initiative, where healthcare workers take the time to explain to patients why antibiotics might not be the best choice for a viral infection, fostering a culture of informed decision-making.

Collaboration is also vital in combating antibiotic resistance. Global initiatives like the Global Antimicrobial Resistance Research and Development Hub aim to facilitate research and development of new antibiotics. This collaborative effort brings together scientists, pharmaceutical companies, and public health experts to share knowledge and resources. The hub focuses on innovation, ensuring that new antibiotics can be developed and introduced while also advocating for responsible use.

And let's not forget the role of education. Programs targeting medical professionals are essential to ensure that doctors are well-informed about the latest guidelines on antibiotic prescribing. Many medical schools around the world now include courses on antimicrobial stewardship, emphasizing the importance of responsible prescribing habits. Picture a group of eager medical students learning not just the science behind antibiotics, but also the ethical responsibility that comes with prescribing them. They leave these courses equipped not just with knowledge, but with a passion for protecting the effectiveness of antibiotics.

Moreover, initiatives like the One Health approach highlight the interconnectedness of human health, animal health, and the environment. By addressing antibiotic use in agriculture, where antibiotics are often overprescribed to livestock, we can significantly reduce resistance. Farmers are increasingly encouraged to adopt sustainable practices, prioritizing animal welfare and minimizing antibiotic use. This shift not only benefits public health but also improves the quality of the food we consume.

On a grassroots level, community-driven initiatives are taking shape, emphasizing the importance of local action. Neighborhood health workshops engage residents in discussions about the consequences of antibiotic misuse. In these workshops, a community leader shares anecdotes from local families affected by resistant infections, sparking empathy and understanding. Such stories encourage participants to reconsider their attitudes toward antibiotics, fostering a sense of responsibility within the community.

The collective efforts of these global initiatives are crucial as we navigate the future of antibiotics. While the challenges ahead may seem daunting, the stories of hope, resilience, and collaboration remind us that we can make a difference. By uniting our voices and actions, we are not only preserving the effectiveness of antibiotics but also ensuring that they remain a viable option for generations to come.

The fight against antibiotic resistance is not a battle fought alone; it is a movement that transcends borders and brings together diverse communities. Through awareness, education, and collaboration, we can create a future where antibiotics continue to be a cornerstone of modern medicine, protecting lives and nurturing health across the globe. As we look ahead, the importance of these global initiatives becomes ever clearer, reminding us that together, we have the power to preserve a vital resource for future generations.

Conclusion

As we draw the curtains on our exploration of penicillin and its journey through the world of medicine, it's essential to reflect on both its remarkable history and the challenges that lie ahead. From the moment Alexander Fleming stumbled upon that moldy Petri dish in 1928, penicillin has transformed the landscape of healthcare. It turned once-fatal infections into manageable conditions, saving countless lives and shaping modern medicine as we know it.

However, with great power comes great responsibility. The rise of antibiotic resistance presents a formidable challenge that threatens to undermine the very foundations of our healthcare systems. It's a sobering thought to consider that something as simple as a bacterial infection, which could once be easily treated with a course of penicillin, might one day become a life-threatening condition again. This reality underscores the importance of not only using antibiotics judiciously but also investing in research and education to combat resistance.

Think back to the stories we've encountered along the way—like that of a mother whose child faced a resistant infection or the community health worker educating families about responsible antibiotic use. These narratives highlight the human element in our fight against antibiotic resistance. Each person impacted by these issues reminds us that we are not just discussing molecules and compounds but lives, families, and futures.

As we look to the future, we must harness the lessons of the past. The innovations in antibiotic research and the global initiatives aimed at preserving antibiotic effectiveness are encouraging. They show that we can unite to tackle this pressing issue. Our combined efforts—through education, responsible prescribing, and community engagement—can turn the tide against resistance.

Moreover, the journey of penicillin teaches us about adaptability and resilience. Just as medicine has evolved, so must our approach to antibiotic stewardship. It's about striking a balance between treating infections effectively and ensuring that these vital drugs remain potent for years to come. By embracing this dual responsibility, we can honor the legacy of penicillin while paving the way for future breakthroughs.

In the end, let's remember that the fight against antibiotic resistance is not just a task for scientists and healthcare professionals; it's a shared responsibility that calls on each of us. Whether it's discussing antibiotic use with your healthcare provider, advocating for responsible practices in your community, or simply educating yourself and others, every action counts.

As we conclude this exploration of penicillin, let's carry forward the hope that through awareness, collaboration, and a commitment to change, we can protect this vital resource. The story of penicillin is far from over; it's an ongoing narrative that invites us all to play a part in shaping a healthier future. Together, we can ensure that the legacy of penicillin continues to be one of healing, innovation, and progress.

Summary of Penicillin's Legacy

Penicillin's legacy is a remarkable tale of discovery, innovation, and the profound impact it has had on human health. From its serendipitous discovery by Alexander Fleming in 1928 to its pivotal role in treating infections during World War II, penicillin has changed the course of medicine in ways that many of us take for granted today. Imagine a time when simple infections could lead to serious illness or even death; penicillin stepped in as a lifesaver, turning this grim reality into a story of hope and recovery.

One of the most touching anecdotes from this era comes from the memories of veterans who received penicillin during the war. Many soldiers who returned home shared stories of how a once-dreaded wound, infected and painful, was healed with the miracle of penicillin. They often expressed gratitude for the advances in medicine that allowed them to return to their families, healthy and whole. These personal narratives highlight not just the scientific triumph but the very human stories behind the numbers and statistics.

Penicillin's impact extends beyond individual stories; it has been a cornerstone in the fight against infectious diseases. Its development led to a new era in medicine, inspiring further research and the discovery of a host of other antibiotics. Yet, as we celebrate its achievements, we must also recognize the challenges that have arisen from its widespread use, notably the rise of antibiotic resistance. This growing concern serves as a reminder that the legacy of penicillin is not just about triumphs but also about the ongoing responsibility we share in using antibiotics wisely.

Looking forward, the legacy of penicillin calls for continued innovation and vigilance. Researchers are tirelessly exploring new avenues to develop antibiotics that can outsmart resistant bacteria, ensuring that future generations will still have effective treatments available. Communities are coming together to advocate for responsible antibiotic use and to educate others about the importance of this issue.

In essence, penicillin's legacy is a tapestry woven with threads of hope, healing, and caution. It serves as a reminder of the power of science to change lives and the importance of our collective efforts to preserve these medical advancements. Every time we take an antibiotic, we are partaking in a story that began nearly a century ago—a story of resilience, healing, and a commitment to a healthier future for all. As we move forward, let us honor this legacy by promoting responsible use and fostering a culture of innovation that will ensure the efficacy of antibiotics for generations to come.

Moving Forward: Lessons Learned from Penicillin

The story of penicillin is more than just a scientific breakthrough; it's a journey filled with valuable lessons that continue to resonate in today's world. When Alexander Fleming accidentally discovered penicillin in 1928, he stumbled upon a game-changing solution to a problem that had plagued humanity for centuries: bacterial infections. This unexpected twist of fate not only saved countless lives but also opened the door to the antibiotic revolution. However, as we reflect on this legacy, it's crucial to consider the lessons learned along the way.

One of the most important takeaways is the significance of responsible antibiotic use. Penicillin's early successes led to its widespread application, but this very success laid the groundwork for a growing problem: antibiotic resistance. As we became more reliant on antibiotics to treat even minor ailments, bacteria evolved and adapted, making some infections increasingly difficult to treat. This highlights the delicate balance we must maintain between utilizing medical advancements and recognizing their limitations. A personal story that captures this lesson involves a family member who developed a stubborn infection that wouldn't respond to the usual antibiotics. It was a frustrating experience, as the options dwindled and anxiety rose. This situation is a reminder of the importance of using antibiotics judiciously and only when necessary.

Another lesson from penicillin's history is the power of collaboration in science. The successful mass production of penicillin during World War II required the combined efforts of scientists, pharmaceutical companies, and government agencies. This spirit of collaboration continues to be essential in today's research landscape. We see it in initiatives that bring together experts from various fields to tackle pressing health issues, from antibiotic resistance to emerging infectious diseases. The COVID-19 pandemic underscored the importance of this collective approach, as scientists worldwide united to develop vaccines at an unprecedented pace.

Moreover, penicillin teaches us the value of continual research and innovation. While penicillin remains a cornerstone of modern medicine, we must not become complacent. The search for new antibiotics and alternative therapies is ongoing, and the lessons from penicillin remind us that the fight against infectious diseases is far from over. This is evident in the efforts of young researchers who share their passion for developing novel treatments, drawing inspiration from the legacy of antibiotics while addressing the challenges of resistance.

Finally, penicillin's story illustrates the need for public awareness and education. Many people today may not fully understand how antibiotics work or the dangers of misusing them. By sharing stories of those affected by antibiotic resistance and the importance of following medical guidance, we can foster a more informed society. For instance, a local community initiative that I came across recently involved a series of workshops on responsible antibiotic use, featuring testimonials from individuals who faced serious health consequences due to antibiotic misuse. These personal narratives resonate deeply, making the message more impactful.

As we move forward, the lessons learned from penicillin serve as a guiding light in our ongoing quest for health and well-being. By embracing responsible use, fostering collaboration, committing to innovation, and raising awareness, we can ensure that the legacy of penicillin continues to benefit future generations.

It's not just about preserving the past; it's about building a healthier, more resilient future for everyone.

Encouraging Responsible Use of Antibiotics

In a world where antibiotics have become a common fixture in our medicine cabinets, it's crucial to remember that these powerful drugs are not a cure-all. Encouraging responsible use of antibiotics is not just a medical guideline; it's a necessity for safeguarding our health and that of future generations.

The tale of antibiotics is fascinating. When penicillin was discovered by Alexander Fleming in 1928, it was hailed as a miracle drug. Suddenly, bacterial infections that once claimed lives were treatable, and people could recover from illnesses that would have been fatal just a few years earlier. However, with great power comes great responsibility. As we became accustomed to using antibiotics for everything from serious infections to minor ailments, the landscape began to shift, giving rise to the very real threat of antibiotic resistance.

Take the story of a close friend of mine, who, after catching a bad cold, insisted on getting antibiotics from her doctor. Despite her physician's advice that antibiotics would not help with her viral infection, she pushed for them, convinced that they would speed up her recovery. Unfortunately, this scenario is not uncommon. Many people believe that antibiotics are a one-size-fits-all solution, unaware that misuse can lead to resistance, rendering these medications ineffective when they are genuinely needed.

So how do we foster a culture of responsible antibiotic use? First, education is key. It's essential for patients to understand what antibiotics are designed for and when they should be used. Antibiotics treat bacterial infections, not viral ones like the flu or common cold. Community awareness campaigns can play a significant role in spreading this message. I recall a local health initiative that brought together doctors, pharmacists, and community leaders to host workshops and informational sessions. They shared real stories, illustrating the consequences of misuse, which helped many in our community grasp the seriousness of the issue.

Moreover, healthcare professionals have a pivotal role in this equation. They must be equipped to educate their patients about the appropriate use of antibiotics. By taking the time to explain why a certain treatment plan is necessary or why antibiotics aren't suitable, they can help demystify these medications. A friend of mine, who is a nurse, often shares anecdotes from her practice, highlighting how a simple conversation about antibiotics can change a patient's perspective. These discussions foster trust and empower patients to make informed decisions about their health.

Another vital aspect is to encourage open communication between patients and healthcare providers. Patients should feel comfortable asking questions and expressing concerns about their treatment options. This dialogue can lead to better understanding and a greater willingness to follow medical advice. When patients grasp the rationale behind a doctor's recommendation, they are more likely to adhere to the prescribed treatment plan.

Lastly, let's not forget about the broader picture. Responsible antibiotic use extends beyond individual patients to include agricultural practices and global health policies. The use of antibiotics in livestock farming can significantly contribute to resistance. By advocating for responsible practices in agriculture, such as reducing unnecessary antibiotic use and promoting animal welfare, we can help combat this pressing issue.

Encouraging responsible use of antibiotics is a collective effort that requires education, communication, and awareness. As we navigate the complexities of modern medicine, let's remember the lessons learned from penicillin's legacy. Each small step—whether it's educating ourselves, talking to our healthcare providers, or advocating for responsible agricultural practices—can contribute to a healthier future where antibiotics remain effective. Together, we can ensure that these precious medications continue to save lives for generations to come.

When it comes to selecting the right antibiotics, particularly penicillin and its derivatives, several additional features deserve attention. These characteristics can significantly impact treatment outcomes, patient experiences, and even public health. Let's delve into some of these features, blending science with real-life stories to highlight their importance.

One key aspect to consider is the route of administration. Antibiotics can be delivered orally, intravenously, or intramuscularly, and the chosen route can affect how quickly and effectively the medication works. For instance, a friend of mine, who battled a severe infection, initially received oral antibiotics. While they were effective, the infection didn't respond as quickly as expected. The doctor decided to switch to an intravenous route, leading to a much faster recovery. This experience highlights how the method of delivery can play a crucial role in treatment success.

Dosage forms also vary widely among antibiotics. Tablets, capsules, liquids, and injections all have their place, depending on the patient's needs. For example, pediatric patients often prefer liquid formulations, which can be easier to swallow. I remember visiting a pharmacy with my niece when she had a throat infection. The pharmacist explained how the liquid form of penicillin was flavored to make it more palatable for children. These thoughtful considerations not only improve compliance but also reduce the stress of taking medication.

Duration of therapy is another essential factor. While it might be tempting to stop taking antibiotics once symptoms improve, it's crucial to complete the full course as prescribed. A neighbor of mine, eager to feel better, stopped her antibiotic regimen early because she felt fine. Unfortunately, the infection returned, and she ended up needing a more potent medication. This incident serves as a valuable reminder of the importance of adhering to treatment plans.

Then there's the impact of antibiotic resistance. As we learn more about how bacteria adapt, the issue of resistance becomes increasingly pressing. My cousin, a microbiologist, often shares anecdotes from her lab work, discussing how some bacteria have become resistant to even the strongest antibiotics. She emphasizes that understanding the mechanisms of resistance is crucial for both medical professionals and patients. We all have a role in combating this challenge, from using antibiotics responsibly to supporting initiatives aimed at reducing resistance.

Side effects and patient tolerance also warrant consideration. While penicillin is generally well-tolerated, some individuals may experience side effects like nausea, diarrhea, or allergic reactions. An acquaintance of mine, excited to start her penicillin treatment for a sinus infection, quickly faced an unpleasant surprise: she developed a rash and other allergic symptoms. It was a learning moment for her, as she realized the importance of informing her healthcare provider about her medical history, including any known allergies. This experience highlights how personal medical histories can shape treatment decisions.

Lastly, it's essential to consider cost and accessibility. The price of antibiotics can vary significantly, affecting patients' access to necessary medications. I recall a community health fair where local pharmacists provided free consultations. They shared stories about patients who skipped doses or didn't fill prescriptions due to cost concerns. The fair highlighted the need for accessible healthcare solutions, ensuring that everyone can afford essential medications.

While penicillin and its derivatives have proven effective for countless infections, it's vital to consider additional features that can influence treatment outcomes. The choices surrounding administration routes, dosage forms, treatment duration, resistance issues, side effects, and cost all play crucial roles in the patient experience. By sharing our stories and supporting one another in understanding these factors, we can navigate the complexities of antibiotic treatment more effectively. Ultimately, informed patients and healthcare providers can work together to ensure the best possible outcomes in the journey toward health.

Case Studies

Penicillin's journey from a serendipitous discovery to a life-saving medication is filled with stories that illuminate its profound impact on modern medicine. Let's explore a few real-life examples that illustrate how penicillin has saved countless lives, each highlighting its transformative power in the face of bacterial infections.

One remarkable story comes from the World War II era. Soldiers were exposed to dangerous infections on the battlefield, and untreated wounds could lead to amputation or even death. Penicillin was introduced as a treatment, and its impact was nothing short of miraculous. I remember reading about a soldier named John, who was injured during a fierce battle. After receiving a severe shrapnel wound, he developed a serious infection. At that time, penicillin was still relatively new, but it was administered to him, and within days, his condition improved significantly. He eventually returned home, alive and healthy, a testament to how this antibiotic could turn the tide in life-or-death situations. This story serves as a reminder of the critical role penicillin played during wartime and how it brought hope to many families.

Another poignant example comes from the life of a young girl named Emily. She was just seven years old when she contracted a severe bacterial infection that led to pneumonia. Her parents were terrified as they watched her struggle to breathe, and the hospital staff worked tirelessly to stabilize her condition. Fortunately, the doctors recognized the seriousness of her illness and quickly started her on a penicillin regimen. Within a week, Emily was sitting up in bed, smiling and chatting with her parents. Her recovery not only saved her life but also brought joy to a family that had been on the brink of despair. This experience highlights how, even in the age of modern medicine, penicillin continues to play a pivotal role in treating infections that can escalate quickly, especially in vulnerable populations like children.

Penicillin has also made a mark in treating infections that arise from less-than-ideal circumstances. For example, consider the case of Mike, a mechanic who cut his hand while working on a rusty vehicle. Initially, he didn't think much of it, but soon, the cut became infected. What started as a minor issue quickly escalated; he developed a fever, and the redness around the wound spread. Realizing the seriousness of the situation, he sought medical help. After a thorough examination, the doctor prescribed penicillin. Within days, Mike noticed a significant improvement, and within a week, the infection was under control. He often shares this story, emphasizing how a simple cut could have turned into a life-threatening situation without the timely use of antibiotics like penicillin.

Then there's the story of Sarah, a woman who faced an unexpected challenge during her pregnancy. After developing a urinary tract infection, her doctor prescribed penicillin to treat it safely. Despite initial concerns about taking antibiotics while pregnant, Sarah's doctor reassured her about the benefits of penicillin, which has a well-established safety record. Thanks to prompt treatment, Sarah's infection cleared up, and she gave birth to a healthy baby girl. Sarah often reflects on how penicillin not only helped her but also provided peace of mind during a critical time in her life. Her experience underscores the importance of accessible healthcare and the vital role antibiotics play in protecting both mothers and their children.

These stories remind us that penicillin is not just a drug but a life-saving tool that has transformed countless lives. From the front lines of war to everyday life, its impact is profound and enduring. Each case represents hope, healing, and the power of medical innovation. As we continue to explore the legacy of penicillin, let's celebrate the countless lives saved and the stories that highlight the importance of responsible antibiotic use, ensuring that this incredible medication remains effective for future generations.

Glossary of Terms

1. Antibiotic: A type of medication that is used to treat bacterial infections. Antibiotics work by killing bacteria or inhibiting their growth.
2. Bacteria: Single-celled microorganisms that can be found in various environments. While many bacteria are harmless or beneficial, some can cause infections and diseases.
3. Penicillin: A group of antibiotics derived from Penicillium fungi. Penicillin was the first antibiotic discovered and is used to treat a variety of bacterial infections.
4. Resistance: The ability of bacteria to withstand the effects of an antibiotic that once killed them or inhibited their growth. This often occurs due to overuse or misuse of antibiotics.
5. Infection: The invasion and multiplication of harmful microorganisms in the body, leading to illness. Infections can be caused by bacteria, viruses, fungi, or parasites.

6. Dosage: The specific amount of medication that should be taken at one time or over a specified period. Proper dosage is crucial for the effectiveness of treatment and to minimize side effects.
7. Administration: The method by which a medication is given to a patient, which can include oral, intravenous (IV), or intramuscular (IM) routes.
8. Therapeutic: Relating to the treatment of a disease or medical condition. Therapeutic antibiotics are used specifically to treat infections.
9. Prophylactic: Referring to measures taken to prevent disease, rather than treating it. Prophylactic antibiotics may be prescribed to prevent infections in certain situations, such as before surgery.
10. Spectrum of Activity: The range of bacteria that an antibiotic is effective against. Antibiotics can have a narrow spectrum (effective against specific types of bacteria) or a broad spectrum (effective against a wide range of bacteria).

11. Culture and Sensitivity Test: A laboratory test used to determine the type of bacteria causing an infection and which antibiotics are effective against it.
12. Pharmacology: The branch of medicine that focuses on the study of drugs, their effects on the body, and their use in treating diseases.
13. Side Effects: Unintended effects that may occur alongside the intended therapeutic effects of a medication. Side effects can range from mild to severe.
14. Adverse Reaction: A harmful or unintended response to a medication, which may require medical intervention.
15. Vaccine: A biological preparation that provides active acquired immunity to a particular infectious disease. Vaccines help prevent infections rather than treat them.
16. Microorganism: A tiny organism, such as bacteria or fungi, that can only be seen with a microscope. Microorganisms include a variety of life forms, some of which are beneficial and others that can cause disease.

17. Efficacy: The ability of a drug to produce a desired effect under ideal conditions, such as clinical trials.
18. Bioavailability: The proportion of a drug that enters the circulation when introduced into the body and is available for therapeutic effect.
19. Toxicity: The degree to which a substance can harm or poison an organism. In the context of antibiotics, toxicity refers to harmful effects on the body.
20. Generic Drug: A medication that is comparable to a brand-name product in dosage form, strength, route of administration, quality, and performance characteristics, but is marketed under its chemical name without the brand label.

This glossary aims to clarify some of the key terms associated with penicillin and antibiotics, making the information more accessible for readers who may not be familiar with medical jargon.

Frequently Asked Questions (FAQ)

1. What is penicillin, and how does it work?

   Penicillin is a type of antibiotic that was first discovered in 1928 by Alexander Fleming. It works by interfering with the ability of bacteria to form cell walls, ultimately leading to their death. This makes it effective against many types of bacterial infections.

2. What types of infections can penicillin treat?

   Penicillin is commonly used to treat a variety of infections, including strep throat, pneumonia, skin infections, and syphilis. However, its effectiveness may vary depending on the specific bacteria involved.

3. Can everyone take penicillin?

   Not everyone can take penicillin. Some individuals may have allergies to penicillin, which can cause reactions ranging from mild rashes to severe anaphylaxis. It's essential to inform your healthcare provider about any known allergies before starting treatment.

4. What are the potential side effects of penicillin?

   Common side effects of penicillin may include

nausea, vomiting, diarrhea, and allergic reactions. More serious side effects, though rare, can include liver damage or severe allergic reactions.

5. Why is there concern about antibiotic resistance?

   Antibiotic resistance occurs when bacteria evolve to become resistant to the effects of medications designed to kill them. This can make infections harder to treat and lead to longer hospital stays, higher medical costs, and increased mortality. The overuse and misuse of antibiotics are significant contributors to this problem.

6. What can I do to prevent antibiotic resistance?

   You can help prevent antibiotic resistance by only using antibiotics when prescribed by a healthcare professional, completing the full course of prescribed antibiotics, and not sharing or using leftover antibiotics. Additionally, practicing good hygiene and vaccination can help prevent infections.

7. Are there alternatives to penicillin if I'm allergic?

   Yes, there are alternative antibiotics available

for individuals with penicillin allergies. Options may include macrolides (like azithromycin) or tetracyclines (like doxycycline), depending on the type of infection and the bacteria involved.

8. How is penicillin produced today?

   Modern penicillin production primarily involves fermentation processes using specific strains of the Penicillium fungus. Advances in biotechnology have improved the efficiency and yield of penicillin production, allowing for better supply and availability.

9. What role does penicillin play in veterinary medicine?

   Penicillin is also used in veterinary medicine to treat bacterial infections in animals. It's crucial in livestock and pet care to help manage infections and promote animal health, though responsible use is necessary to prevent resistance.

10. What is being done to combat antibiotic resistance?

    Global initiatives are underway to combat antibiotic resistance through improved surveillance, research into new antibiotics, public education campaigns, and stricter regulations on antibiotic use in both humans and animals. Collaborative efforts between healthcare providers, policymakers, and the public are vital to address this ongoing challenge.

These FAQs aim to clarify some common queries about penicillin, its uses, and the critical issue of antibiotic resistance. If you have more specific questions or concerns, it's always best to consult a healthcare professional.

www.ingramcontent.com/pod-product-compliance
Lightning Source LLC
Chambersburg PA
CBHW052147220526
45471CB00004B/1556